Remote Patient Monitoring: A Computational Perspective in Healthcare

Published 2023 by River Publishers
River Publishers
Alsbjergvej 10, 9260 Gistrup, Denmark
www.riverpublishers.com

Distributed exclusively by Routledge
605 Third Avenue, New York, NY 10017, USA
4 Park Square, Milton Park, Abingdon, Oxon OX14 4RN

Remote Patient Monitoring: A Computational Perspective in Healthcare / by
Rishabha Malviya, Priyanshi Goyal.

Routledge is an imprint of the Taylor & Francis Group, an informa
business

ISBN 978-87-7004-025-9 (paperback)

ISBN 978-10-0381-024-7 (online)

ISBN 978-1-032-62799-1 (ebook master)

A Publication in the River Publishers series
RAPIDS SERIES IN BIOTECHNOLOGY AND MEDICAL RESEARCH

Remote Patient Monitoring: A Computational Perspective in Healthcare

Rishabha Malviya

Department of Pharmacy, School of Medical and Allied Sciences,
Galgotias University, Greater Noida, India

Priyanshi Goyal

Department of Pharmacy, School of Medical and Allied Sciences,
Galgotias University, Greater Noida, India

River Publishers

Routledge
Taylor & Francis Group

NEW YORK AND LONDON

Contents

Preface

Healthcare is an area where technology and services are evolving quickly. Remote patient monitoring is a recent concept in this field that has many benefits in an increasing world population with increasing health issues. With relatively simple applications to monitor patients inside hospital rooms, the technology has developed to the extent that the patient can be allowed normal daily activities at home while still being monitored with the use of modern communication and sensor technologies. An amazing impact on global critical care has resulted from the rising usage of mobile technology and smart gadgets in the health sector. In clinical settings, these technologies are being used by specialists and clinicians to bring about fundamental shifts in the delivery of medical care. Many people are also benefiting from E-Health (social insurance supported by ICT) and M-Health (mobile health applications) in order to improve their health. The Internet of Things is making it easier to link devices that are Internet-ready so that doctors can receive real-time updates on their patients' conditions.

Since it enables healthcare professionals to continuously monitor patients outside of the usual context, remote patient monitoring (RPM) has emerged as a crucial tool in the healthcare sector. It reduces the number of in-person appointments to medical facilities or hospitals and conserves the time and energy required for recovery. RPM tools promote openness and transparency. Additionally, it offers a deeper comprehension of ailments and therapies, enabling consumers to have greater influence over their healthcare strategies. The standard of care is often improved when the patient and the healthcare professional are both more informed. This book provides a review of the recent advances in remote healthcare and monitoring in both with-contact and contactless methods. This book discusses the potential scope of applicability of artificial intelligence methods within the telehealth domain. This premier reference source is an essential resource for hospital administrators, medical technicians, healthcare professionals, medical students and educators, librarians, researchers, and academicians. These breakthroughs make it possible to practice medicine any time, any place, and on any device.

About the Authors

Dr. Rishabha Malviya completed B. Pharmacy from Uttar Pradesh Technical University and M. Pharmacy (Pharmaceutics) from Gautam Buddha Technical University, Lucknow Uttar Pradesh. His PhD (Pharmacy) work was in the area of Novel formulation development techniques. He has 12 years of research experience and presently working as Associate Professor in the Department of Pharmacy, School of Medical and Allied Sciences, Galgotias University since past 8 years. His area of interest includes formulation optimization, nanoformulation, targeted drug delivery, localized drug delivery and characterization of natural polymers as pharmaceutical excipients. He has authored more than 150 research/review papers for national/international journals of repute. He has 58 patents (19 grants, 38 published, 1 filed) and publications in reputed National and International journals with total of 191 cumulative impact factor. He has also received an Outstanding Reviewer award from Elsevier. He has authored/edited/editing 46 books (Wiley, CRC Press/Taylor and Francis, IOP publishing. River Publisher Denmark, Springer Nature, Apple Academic Press/Taylor and Francis, Walter de Gruyter, and OMICS publication) and authored 31 book chapters. His name has included in word's top 2% scientist list for the year 2020 and 2021 by Elsevier BV and Stanford University. He is Reviewer/Editor/Editorial board member of more than 50 national and international journals of repute. He has invited as author for "Atlas of Science" and pharma magazine dealing with industry (B2B) "Ingredient south Asia Magazines".

Priyanshi Goyal completed 10th and 12th from Krishna International school, Aligarh affiliated by CBSE. B.Pharm from Aligarh college of pharmacy, Aligarh affiliated by AKTU, Lucknow. She is pursuing M. Pharm from Galgotias University, Greater Noida. She has authored 1 chapter in Springer Nature (In Press). She the author of 1 bookwith Apple Academic Press, Taylor and Francis Group.

Future of Remote Patient Monitoring with AI: An Overview

Abstract

Artificial intelligence (AI) is becoming increasingly popular in the medical field. One particular piece of healthcare software that is frequently used, remote patient monitoring (RPM), helps doctors keep tabs on patients who aren't in the same physical space as them, whether they're at home with an elderly loved one or in a hospital. Manual patient monitoring systems are only as trustworthy as the time management of the staff, which in turn is based on the amount of work they have to do. Invasive methods that require skin contact are typically used in the medical practice of patient monitoring. The purpose of this research is to research the effects of artificial intelligence on RPM, and the difficulties and emerging tendencies associated with this field. In this overview, we discuss some of the difficulties and emerging tendencies in applying AI to RPM systems. Based on the trends and obstacles, potential uses of AI in RPM in the future are discussed.

1.1 Introduction

Remote patient monitoring, which uses flexible sensors to assist clinicians in several surgical and medical wards of a general hospital, is growing rapidly [1–3]. Telemedicine, wearables, and contact-based sensors are used in healthcare to achieve this [4–6]. RPM helps diagnose and treat mental and movement issues by measuring physiological signs including heart rate and motion detection [7, 8]. AI systems help define and predict diseases by analyzing medical images and correlating symptoms, and molecular markers [9, 10]. Because AI has the potential to improve healthcare service delivery, clinicians are studying a wide range of issues related to disease risk evaluation, ongoing patient care, and the application of AI to reduce unfavorable outcomes [11]. AI speeds up genome sequencing and the development of new drugs and cures based on insights that were previously unobservable due to data complexity. Using specialized algorithms, machine learning—a type of artificial intelligence—may help therapists quickly understand complex data [12, 13]. They can classify a patient's motion or activities, predict health decline, and more [14, 15]. These AI systems can process vast datasets and learn complex patterns for decision-making [16]. Computing power has increased dramatically, enabling deep learning and artificial neural network technologies to manage and optimize exceedingly complex information [17, 18]. A centralized hub for monitoring and administering an Internet of Things architecture can automate many boring tasks. This reduces medical errors and improves patient safety [19]. Traditional applications of RPM have focused on telehealth monitoring of patients in faraway locations, home monitoring of chronically ill patients using wearable devices or sensors, and aged care monitoring using video surveillance in their own homes via wireless body sensors, but its non-intrusive nature makes it an attractive option for use in hospitals by patients recovering from surgery and those in intensive care. Digital technologies that don't interfere with patients' daily lives can improve patient monitoring. As shown in Figure 1.1, machine

Figure 1.1: Artificial Intelligence enables RPM.

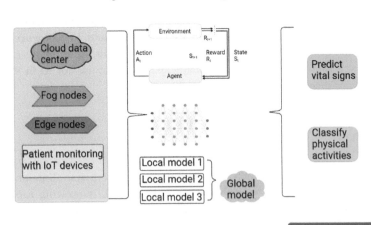

learning (ML) and AI can help healthcare providers visualize patients' health states based on vital signs and activity detection. These apps can display data to help doctors diagnose and predict patients' health. We wrote this review because AI and ML could transform healthcare. This article examines modern RPM system noninvasive approaches. AI affects RPM's early health decline detection, personalized monitoring, and adaptive learning. Finally, the barriers to AI/ML-based remote healthcare monitoring and current initiatives to overcome them are discussed.

1.2 Impact of AI on RPM

1.2.1 Recognizing patient worsening early

Early vital sign deterioration detection in critically ill hospital patients is crucial to avert clinical worsening. Traditional patient monitoring reports patient vital signs and clinical state. Core body temperature, heart rate, respiration rate, and mean arterial pressures (MAPs) are continuous predictors for emergency department patients [20]. New patient monitoring algorithms assess several physiological features. Thus, an indicator evaluates a health risk or physiological instability [21]. In a case series, Posthuma et al. [22] showed that wireless remote vital sign monitoring devices in surgical wards could speed recovery by recognizing worsening patients. The nursing staff in this study found the systems reasonably helpful, but clinical judgment was needed to evaluate the patient. They noted that there are no rules or criteria for using these systems, thus medical practitioners should continue to make their own decisions regarding the best solution for their practices. Kellett and Sebat [23] discussed vital sign recording after emphasizing its importance. Despite manual, intermittent surveillance in most hospital wards, researchers found no consensus on the minimal time between vital sign recordings. Kellett also stressed the importance of patient monitoring to detect anomalies. Early warning scores (EWS), a manually calculated screening metric, can evaluate clinical techniques to predict patient decline [24, 25]. EWS systems are sporadic, which limits their availability [26]. Patients could only undergo continuous vital sign monitoring in the ER and NICU before low-patient-to-staff ratio ICUs. Alshwaheen et al. [27] presented LSTM-RNNs to forecast patient decline in intensive care units. The algorithm predicted degradation one hour in advance and outperformed the conventional approach in classification. Muralitharan et al. [28] extended this work and showed that machine learning-based EWS may be utilized in a range of acute general medical and surgical wards, including ambulatory and home care settings, and outperform manual techniques in performance and accuracy. Despite several research predicting health outcomes, da Silva et al. [29] predicted future vital sign deterioration using RNNs and LSTM to historical data from electronic medical records. These expected vital signs were utilized to diagnose a deteriorating health state early using a clinical prognostic approach with 80% accuracy. Clinicians need transparency and explainability to embrace AI models. Scientists developed xAI-EWS, an explainable AI early warning system, in 2020, to detect severe illnesses [30]. A DTD explanation module and a TCN prediction module were used to create the xAI-EWS. Clinical specialists employed the three most common emergency room visits—AKI, ALI, and septic shock—to test the system (ALI). Clinicians gained confidence in the model's prediction ability by learning about its internal workings without having to grasp the processes.

1.2.2 Individualized surveillance

Due to this lack of attention to individual variation, standard medical practice makes broad generalizations about a patient's illness and prescribes a uniform treatment plan regardless of how each patient responds to the drug or proce-dure [31]. RPM architecture collects patient data for AI modeling using the Internet of Things and cloud computing. Modern healthcare systems require patient-centered or personalized monitoring, especially for long-term diseases like mental health, diabetes, and others. Distributed networks like fog and edge computing enable personalize monitoring of a patient's Internet of Things (IoT) devices. Mukherjee et al. [32] proposed a cloud edge-fog framework for personalized health monitoring to predict patient mobility and recommend local medical facilities in emergencies. IoT platform data must be removed from devices and integrated on a cloud server for data analytics. This concerns patient health data security and privacy. It also consumes much of the tech-nology and power. Google's federated learning architecture for AI approaches might train a single AI model across numerous distributed edge devices using local patient data without data exchange or integration. A cloud server receives all local model weights. The model weights construct a powerful, universal AI model. This decentralized method always stores patient data on their device. The robust global model can be used to improve local data classification and prediction. The scientific community has embraced IoT applications. Zheng et al. proposed a healthcare IoMT-centric federated transfer learning system in 2021 [33]. Nguyen et al. [32] discussed healthcare federated learning frame-works, their benefits, needs, applications, trends, and challenges. FedHome, a cloud-based federated learning architecture, was used to train local models for in-home health monitoring [35]. That study built a generative convolutional autoencoder (GCAE) to analyze uneven and nonidentical distribution data and track individual health accurately. The suggested strategy outperforms baseline models with 95.87% for balanced data and 95.41% for unbalanced data. FedHealth classified physical activity for personalized monitoring by Chen et al. [36]. Data aggregation and transfer learning produce personalized models in the study's federated learning system. Two classification issues assessed FedHealth. One is to classify physical activities into predetermined categories 99.4% more reliably than baseline models. The second one classified Parkinson's disease patients' postural tremor and drooping arms with 84.3% and 74.9% accuracy. Researchers [7, 8] have presented a heterogeneous FedStack architecture to support diverse architectural local models on the patient to categorize patients' physical activity and generate a valid global model utilizing local model predictions.

1.2.3 Adaptive learning

Reward-driven reinforcement learning can design and execute an action plan. Machine learning teaches goal-setting in uncertain, complicated environments. It can learn from mistakes by trying new solutions and obtaining positive or negative feedback [37]. A reinforcement learning agent has no background data or information. Agents discover patterns via experience. As shown in Figure 1.2, an action (A_t) is conducted to transition from the current state (S_t) to $S_t + 1$ at time t + 1 time steps later. These behaviors are rewarded beforehand. If the agent follows policy, they get (R_t). If not, they get fined. Researchers have applied reinforcement learning to dynamic therapeutic regimes for chronic, mental, and infectious diseases [38]. In these cases, decision rules dictate therapy, drug dose, and re-examination recommendations. Yu et al. explored RL in healthcare in 2023 [37]. Reinforcement-based treatments for psychiatric illnesses, cancer, diabetes, anemia, HIV, and chronic diseases have been researched. Researchers found that releasing a patient from the hospital requires a physician (learning agent) to ventilate and monitor the patient's environment [39]. This paper details healthcare reinforcement learning. Deep reinforcement learning and real-time wearable sensor data were used to prescribe medicine timing and dosage [40]. An intelligent system employs deep learning and reinforcement learning algorithms to maximize patient medication compliance [41]. Just-in-time adaptive interventions (JITAIs) demand immediate action to provide patients with the right care at the right moment. Adaptive learning can help attain this goal [42]. Wang, et al. [43] optimized mobile healthcare intervention methods using data-driven reinforcement learning. A reinforcement learning

Figure 1.2: Adaptive learning mechanism.

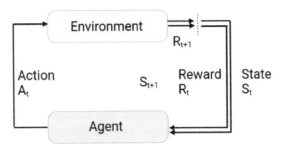

technique can personalize digital adaptive therapies like cellphone notifications to help patients manage their health [44]. Opportunity identification and
intervention selection were used. The intervention selection model dictates
intervention type and frequency. This procedure seeks the right opportunity. AI
can change healthcare applications employing cutting-edge mechanisms such as
RL and federated learning.

1.3 AI in RPM: Challenges and Trends

A technology-supported patient monitoring system requires medical personnel
dedication and input. In 2021, Ede et al. conducted a qualitative study on staff
expectation of wireless, noncontact vital sign monitoring, staff views of its use
in the intensive care unit (ICU), patients' reactions, and staff attitudes about
its introduction [45]. Nine nurses with a median of two years of ICU experience
were interviewed on five topics: ICU staff should expect constant monitoring,
problem-solving, a hierarchy of monitoring, and a culture of trust; patient and
relative monitoring experiences in relation to current wired monitoring and
expectations of noncontact monitoring; and staff perceptions of these experiences. AI can analyze, predict, and classify data, but healthcare is resistant to
embracing it [46]. We examine the problems of using AI in remote monitoring
systems for reliable vital sign readings and activity recognition. The initiatives
address these issues.

1.3.1 Capacity to explain using AI or ML

Humans' inability to comprehend the AI/ML model output is the biggest challenge. ML models beat humans in data interpretation and outcome prediction,

but they cannot explain their reasoning or detect algorithm failures. This is a major barrier to AI and machine learning in healthcare [47]. Most ML models—neural networks (NNs), SVMs, and others—are black boxes. These models cannot elaborate findings or show causal links between independent and dependent variables [48]. AI and ML can only be adaptable if they offer medically understandable outputs and structures [49].

License sensitivity may reveal a trained neural network's input–output relationships. Tree-based methods don't need parametric input and output variables. Interpretable structures help clinical decision-making. Classification and regression trees demonstrate this [49, 50]. Pattern recognition, naive Bayes, heuristics, regression, fuzzy sets, and trees all provide interpretable structures [51]. The cooperative game theory-based Shapley additive explanations (SHAP) [52] can improve AI model transparency and explainability [53]. Shapley values indicate each input characteristic's importance to the forecast in this method. The prediction model's Shapley values provided a high-level overview and detailed views of each feature. Linardatos et al. studied machine learning interpretability methods in 2020 [54]. The study's findings on deep learning model interpretability vary. It includes DeepLIFT, deconvolution, guided back-propagation, gradients explanation, and gradient-weighted class activation mapping. A paradigm for precise and effective personal healthcare was put up by Raza et al. in 2022, employing federated transfer learning and an explainable AI (ExAI) model for EEG signal classification [55]. Each patient's correct and efficient healthcare was planned by a researcher [55]. Khodaban-dehloo et al. [56] created HealthXAI, an adaptable AI system that detects early indicators of linked home weakening. The anomaly feature vector determined unusual behavior severity. The researchers constructed a dashboard for medical practitioners to instantly access anomaly data, scores, and plain-language machine-generated descriptions.

1.3.2 Privacy

The opaque nature of deep neural networks makes it impossible to predict their input data-based conclusions. They may unintentionally find biased user data patterns. Data leakage is high. Iwasawa et al. [57] assessed classical deep learning network attributes using wearable data, confirming the occurrence. A simple logistic regressor trained using convolutional neural network (CNN) features from low-dimensional activity data had 84.7% user categorization accuracy. Raw sensor data categorization accuracy was 35.2% for the same classifier. This shows how a human activity identification deep learning network could leak private data [58]. Data perturbation and transformation are suggested to overcome privacy concerns with machine learning algorithms [59]. Adversarial neural networks were recommended to reduce user identification false positives. Unlike other methods, adversarial loss can only conceal sensitive data like a user's gender and personal information. To protect user identification, raw sensor data were style and content analyzed. Raw sensor data can be turned into random noises with the same "content" [58]. Gati et al. [60] introduced a deep private auto-encoder (dPA) to guarantee data perturbation-based differential privacy. A differential privacy preserving the softmax layer was added to feature extraction layers to safeguard privacy during classification and prediction. Academics are addressing privacy with blockchain technology. A researcher developed a blockchain-based architecture for e-health apps to protect user privacy due to the distributed ledger's immutability and anonymity [61]. Ul Hassan et al. [62] designed differential privacy for the blockchain to prevent data loss from perturbations. These writers created all blockchain layers for differentiated privacy. Federated learning, another AI approach, is popular because it allows collaborative learning while respecting user privacy. By exchanging only AI model parameters, federated learning secures sensitive data. Singh et al. [63] presented a blockchain–federated learning architecture for smart healthcare patient privacy. Federated learning requires the researcher to submit model parameters to the cloud. Data disruption, blockchain, and federated learning are helping healthcare apps handle patient privacy and data leaks.

1.3.3 Uncertainty

Discoveries from modeling, data collection, and deep neural network construction complicate healthcare AI adoption [64]. RPM systems need data. Human error, measurement instrument noise, and real-world complexity cause uncertainty. The data will be utilized to build and train a deep neural network (DNN) model, but DNN's many hyperparameters create ambiguity in model structure and training. Variability in input data (aleatoric uncertainty) and output models (epistemic uncertainty) causes modeling uncertainty [65]. Uncertainty quantification (UQ) reduces unknowns, helping optimization and decision-making. Abdar et al. [66] reviewed the literature on ML and DL model uncertainty. This study discusses ensemble and Bayesian strategies for resolving deep learning models' interpretability and dependability issues, including BDL [67, 68]. UQ is also examined in machine-learning-supported medical decision-making [69]. The researchers identified four interrelated UQ issues, including healthcare deep learning models. Healthcare research lacks theory, hence there is no quantitative model to draw results. Second, casual models aren't available because DL models can only infer so many conclusions. Quantifying uncertainty with high sensitivity to imprecise real-world data. Finally, training and re-computing/re-evaluating in deep learning add significant compute expenses.

1.4 Conclusion

Information systems and artificial intelligence have transformed healthcare applications. Over the past decade, tracking patients' vital signs and physical activity has improved, providing a more reliable health assessment. Data modeling and transmission advances have enabled RPM systems to learn from patients' habits, adapt to their preferences through patient-centric applications, and forecast health decline. This article demonstrates how AI may improve RPMs by learning, predicting, and classifying patient behavior and vital signs. AI is studied for health, activity, disease, and emergency response. Federated learning allows a patient-centric monitoring system that prioritizes individual needs without compromising privacy. Reinforcement learning helps RPMs adjust and learn patient behavior. The evidence-based implications of such cutting-edge AI methods on RPM systems are examined. AI must overcome explainability, privacy, and uncertainty to revolutionize RPM services. Data learning also faces feature extraction, imbalanced labelling, enormous data volumes, and complex data processing. We examine RPM AI's challenges in this study. Despite using RPM systems with physiological monitoring and movement tracking, the study neglected EEG monitoring and nervous system diseases. Not all chronic disease monitoring studies were analyzed. Health care apps can transcend study limits and obstacles by integrating cutting-edge technologies like the Cloud, Edge, Fog, and Blockchain, as well as artificial intelligence methods like federated learning and reinforcement learning. Supervised and unsupervised learning have yielded cutting-edge AI results. However, customized and predictive patient monitoring and increased medical staff resources are needed to modernize healthcare.

References

[1]   Joshi, M., Archer, S., Morbi, A., Arora, S., Kwasnicki, R., Ashrafian, H., Khan, S., Cooke, G., and Darzi, A. (2021) Short-term wearable sensors for in-hospital medical and surgical patients: Mixed methods analysis of patient perspectives. *JMIR Perioperative Medicine* 4(1), e18836. https://doi.org/10.2196/18836

[2]   Liu, H., Wang, L., Lin, G., and Feng, Y. (2022) Recent progress in the fabrication of flexible materials for wearable sensors. *Biomaterials Science* 10(3), 614–632. https://doi.org/10.1039/d1bm01136g

[3]   Weenk, M., Bredie, S., Koeneman, M., Hesselink, G., van Goor, H., and van de Belt, T. H. (2020) Continuous monitoring of vital signs in the general ward using wearable devices: Randomized controlled trial. *Journal of Medical Internet Research* 22(6), e15471. https://doi.org/10. 2196/15471

[4]   Heijmans, M., Habets, J., Kuijf, M., Kubben, P., and Herff, C. (2019) Evaluation of Parkinson's disease at home: Predicting tremor from wearable sensors. *In 2019 41st Annual International Conference of the IEEE Engineering in Medicine and Biology Society (EMBC)*, IEEE. https://doi.org/10.1109/embc.2019.8857717

[5]   Dias, D. and Cunha, J. P. S. (2018) Wearable health devices—Vital sign monitoring, systems and technologies. *Sensors* 18(8), 2414. https://doi.org/10.3390/s18082414

[6]   Malasinghe, L. P., Ramzan, N., and Dahal, K. (2017) Remote patient monitoring: A comprehensive study. *Journal of Ambient Intelligence and Humanized Computing* 10(1), 57–76. https://doi.org/10.1007/s12652-017-0598-x

[7]   Shaik, T., Tao, X., Higgins, N., Gururajan, R., Li, Y., Zhou, X., and Acharya, U. R. (2022) Fedstack: Personalized activity monitoring using stacked federated learning. *Knowledge-Based Systems* 257(12), 109929. https://doi.org/10.1016/j.knosys.2022.109929

[8]   Shaik, T., Tao, X., Higgins, N., Xie, H., Gururajan, R., and Zhou, X. (2022) AI enabled RPM for mental health facility. In *Proceedings of the 1st ACM Workshop on Mobile and Wireless Sensing for Smart Healthcare,* Association for Computing Machinery, 26–32. https://doi.org/10.1145/3556551.3561191

[9]   Miller, D. D. and Brown, E. W. (2018) Artificial intelligence in medical practice: The question to the answer? *The American Journal of Medicine* 131(2), 129–133. https://doi.org/10.1016/j.amjmed.2017.10.035

[10]   Schnyer, D. M., Clasen, P. C., Gonzalez, C., and Beevers, C. G. (2017) Evaluating the diagnostic utility of applying a machine learning algorithm to diffusion tensor MRI measures in individuals with major depressive disorder. *Psychiatry Research: Neuroimaging* 264, 1–9. https://doi.org/10.1016/j.pscychresns.2017.03.003

[11]   Torous, J., Nicholas, J., Larsen, M., Firth, J., and Christensen, H. (2018) Clinical review of user engagement with mental health smartphone apps: Evidence, theory and improvements. *Evidence-Based Mental Health* 21(3), 116–119. https://doi.org/10.1136/eb-2018-102891

[12]   Helm, J. M., Swiergosz, A. M., Haeberle, H. S., Karnuta, J. M., Schaffer, J. L., Krebs, V. E., Spitzer, A. I., and Ramkumar, P. N. (2020) Machine learning and artificial intelligence: Definitions, applications, and future directions. *Current Reviews in Musculoskeletal Medicine* 13(1), 69–76. https://doi.org/10.1007/s12178-020-09600-8

[13]   Krittanawong, C., Johnson, K. W., Choi, E., Kaplin, S., Venner, E., Muru-
gan, M., Wang, Z., Glicksberg, B. S., Amos, C. I., Schatz, M. C., and Tang, W.
W. (2022) Artificial intelligence and cardiovascular genetics. *Life* 12(2),
279. https://doi.org/10.3390/ life12020279

[14]   Liu, Z., Zhu, T., Wang, J., Zheng, Z., Li, Y., Li, J., and Lai, Y.
(2022) Functionalized fiber-based strain sensors: Pathway to next
generation wearable electronics. *Nano-Micro Letters* 14(1), 1–39.
https://doi.org/10.1007/s40820-022-00806-8

[15]   Huang, C., Fukushi, K., Wang, Z., Nihey, F., Kajitani, H., and Nakahara,
K. (2022) Method for estimating temporal gait parameters concerning
bilateral lower limbs of healthy subjects using a single in-shoe motion
sensor through a gait event detection approach. *Sensors* 22(1), 251.
https://doi.org/10.3390/s22010351

[16]   Dean, N. C., Vines, C. G., Carr, J. R., Rubin, J. G., Webb, B. J., Jacobs, J.
R., Butler, A. M., Lee, J., Jephson, A. R., Jenson, N., Walker, M., Brown,
S. M., Irvin, J. A., Lungren, M. P., and Allen, T. L. (2022) A pragmatic,
stepped wedge, cluster-controlled clinical trial of real-time pneumonia
clinical decision support. *American Journal of Respiratory and Critical
Care Medicine* 205(11), 1330–1336. https://doi.org/10. 1164/rccm.202109-
2092OC

[17]   Bini, S. A. (2018) Artificial intelligence, machine learning, deep learning, and cognitive computing: What do these terms mean and how will
they impact health care? *The Journal of Arthroplasty* 33(8), 2358–2361.
https://doi.org/10.1016/j.arth.2018.02.067

[18]   Kalfa, D., Agrawal, S., Goldshtrom, N., LaPar, D., and Bacha, E. (2020)
Wireless monitoring and artificial intelligence: A bright future in car-
diothoracic surgery. *The Journal of Thoracic and Cardiovascular Surgery*
160(3), 809–812. https://doi.org/10.1016/j.jtcvs.2019.08.141

[19]   Tandel, S., Godbole, P., Malgaonkar, M., Gaikwad, R., and Padaya, R.
(2022) An improved health monitoring system using iot. *SSRN* 4109039.
https://doi.org/10.2139/ssrn.4109039

[20]   Asiimwe, S. B., Vittinghoff, E., and Whooley, M. (2020) Vital signs
data and probability of hospitalization, transfer to another facility,
or emergency department death among adults presenting for medical
illnesses to the emergency department at a large urban hospital in
the United States. *The Journal of Emergency Medicine* 58(4), 570–580.
https://doi.org/10.1016/j.jemermed.2019.11.020

[21]   Helman, S., Terry, M. A., Pellathy, T., Williams, A., Dubrawski, A., Cler-
mont, G., Pinsky, M. R., Al-Zaiti, S., and Hravnak, M. (2022) Engaging
clinicians early during the development of a graphical user display of
an intelligent alerting system at the bedside. *International Journal of
Medical Informatics* 159, 104643. https://doi.org/10.1016/j.ijmedinf.2021.
104643

[22] Posthuma, L., Downey, C., Visscher, M., Ghazali, D., Joshi, M., Ashrafian, H., Khan, S., Darzi, A., Goldstone, J., and Preckel, B. (2020) Remote wireless vital signs monitoring on the ward for early detection of deteriorating patients: A case series. *International Journal of Nursing Studies* 104, 103515. https://doi.org/10.1016/j.ijnurstu.2019.103515

[23] Kellett, J. and Sebat, F. (2017) Make vital signs great again—A call for action. *European Journal of Internal Medicine* 45, 13–19. https://doi.org/10.1016/j.ejim.2017.09.018

[24] Garca-del Valle, S., Arnal-Velasco, D., Molina-Mendoza, R., and Gomez-Arnau, J. I. (2021) Update on early warning scores. *Best Practice & Research Clinical Anaesthesiology* 35(1), 105–113. https://doi.org/10.1016/j.bpa.2020.12.013

[25] Vinegar, M. and Kwong, M. (2021) Taking score of early warning scores. *University of Western Ontario Medical Journal* 89(2). https://doi.org/10.5206/uwomj.v89i2.10518

[26] Downey, C., Chapman, S., Randell, R., Brown, J., and Jayne, D. (2018) The impact of continuous versus intermittent vital signs monitoring in hospitals: A systematic review and narrative synthesis. *International Journal of Nursing Studies* 84, 19–27. https://doi.org/10.1016/j. ijnurstu.2018.04.013

[27] Alshwaheen, T. I., Hau, Y. W., Ass'Ad, N., and Abualsamen, M. M. (2021) A novel and reliable framework of patient deterioration prediction in intensive care unit based on long short-term memory-recurrent neural network. *IEEE Access* 9, 3894–3918. https://doi.org/10.1109/ACCESS.2020.3047186

[28] Muralitharan, S., Nelson, W., Di, S., McGillion, M., Devereaux, P., Barr, N. G., and Petch, J. (2020) Machine learning-based early warning systems for clinical deterioration: Systematic scoping review *Journal of Medical Internet Research* 23(2), e25187 https://doi.org/10.2196/25187

[29] da Silva, D. B., Schmidt, D., da Costa, C. A., da Rosa Righi, R., and Eskofier, B. (2021) Deepsigns: A predictive model based on deep learning for the early detection of patient health deterioration. *Expert Systems with Applications* 165, 113905. https://doi.org/10.1016/j.eswa.2020. 113905

[30] Lauritsen, S. M., Kristensen, M., Olsen, M. V., Larsen, M. S., Lauritsen, K. M., Jørgensen, M. J., Lange, J., and Thiesson, B. (2020) Explainable artificial intelligence model to predict acute critical illness from electronic health records. *Nature Communications* 11(1), 3852. https://doi.org/10.1038/s41467-020-17431-x

[31] Chen, G., Xiao, X., Zhao, X., Tat, T., Bick, M., and Chen, J. (2021) Electronic textiles for wearable point-of-care systems. *Chemical Reviews* 122(3), 3259–3291. https://doi.org/10.1021/acs.chemrev.1c00502

[32] Mukherjee, A., Ghosh, S., Behere, A., Ghosh, S. K., and Buyya, R. (2020) Internet of health things (IoHT) for personalized health care using

integrated edge-fog-cloud network. *Journal of Ambient Intelligence and Humanized Computing* 12(1), 943–959. https://doi.org/10.1007/ s12652-020-02113-9

[33] Zheng, X., Shah, S. B. H., Ren, X., Li, F., Nawaf, L., Chakraborty, C., and Fayaz, M. (2021) Mobile edge computing enabled efficient communication based on federated learning in internet of medical things. *Wireless Communications and Mobile Computing* 2021, 1–10. https:// doi.org/10.1155/2021/4410894

[34] Nguyen, D. C., Pham, Q.-V., Pathirana, P. N., Ding, M., Seneviratne, A., Lin, Z., Dobre, O., and Hwang, W.-J. (2023) Federated learning for smart healthcare: A survey. *ACM Computing Surveys* 55(3), 1–37. https://doi.org/10.1145/3501296

[35] Wu, Q., Chen, X., Zhou, Z., and Zhang, J. (2022) FedHome: Cloud-edge based personalized federated learning for in-home health monitoring. *IEEE Transactions on Mobile Computing* 21(8), 2818–2832. https://doi.org/10.1109/tmc.2020.3045266

[36] Chen, Y., Qin, X., Wang, J., Yu, C., and Gao, W. (2020) FedHealth: A federated transfer learning framework for wearable healthcare. *IEEE Intelligent Systems* 35(4), 83–93. https://doi.org/10.1109/mis.2020.2988604

[37] Yu, C., Liu, J., Nemati, S., and Yin, G. (2023) Reinforcement learning in healthcare: A survey. *ACM Computing Surveys* 55(1), 1–36. https:// doi.org/10.1145/3477600

[38] Laber, E. B., Lizotte, D. J., Qian, M., Pelham, W. E., and Murphy, S. A. (2014) Dynamic treatment regimes: Technical challenges and applications. *Electronic Journal of Statistics* 8(1), 1225–1272. https://doi.org/10.1214/14-ejs92

[39] Chen, I. Y., Joshi, S., Ghassemi, M., and Ranganath, R. (2021) Probabilistic machine learning for healthcare. *Annual Review of Biomedical Data Science* 4(1), 393–415. https://doi.org/10.1146/annurev-biodatasci-092820-033938

[40] Watts, J., Khojandi, A., Vasudevan, R., and Ramdhani, R. (2020) Optimizing individualized treatment planning for parkinson's disease using deep reinforcement learning. In *2020 42nd Annual International Conference of the IEEE Engineering in Medicine & Biology Society (EMBC)*. IEEE. https://doi.org/10.1109/embc44109.2020.9175311

[41] Naeem, M., Paragliola, G., and Coronato, A. (2021) A reinforcement learning and deep learning based intelligent system for the support of impaired patients in home treatment. *Expert Systems with Applications* 168, 114285. https://doi.org/10.1016/j.eswa.2020.114285

[42] Nahum-Shani, I., Smith, S. N., Spring, B. J., Collins, L. M., Witkiewitz, K., Tewari, A., and Murphy, S. A. (2017) Just-in-time adaptive interventions (JITAIs) in mobile health: Key components and design principles for

ongoing health behavior support. *Annals of Behavioral Medicine* 52(6), 446–462. https://doi.org/10.1007/s12160-016-98

[43]   Wang, S., Zhang, C., Kröse, B., and van Hoof, H. (2021) Optimizing adaptive notifications in Mobile health interventions systems: Reinforcement learning from a data-driven behavioral simulator. *Journal of Medical Systems* 45(12), 1–8. https://doi.org/10.1007/s10916-021- 01773-0

[44]   Gönül, S., Namlı, T., Cosÿar, A., and Toroslu, _ I. H. (2021) A reinforcement learning based algorithm for personalization of digital, just-in-time, adaptive interventions. *Artificial Intelligence in Medicine* 115(0933-3657), 102062. https://doi.org/10.1016/j.artmed.2021.102062

[45]   Ede, J., Vollam, S., Darbyshire, J. L., Gibson, O., Tarassenko, L., and Watkinson, P. (2021) Non-contact vital sign monitoring of patients in an intensive care unit: A human factors analysis of staff expectations. *Applied Ergonomics* 90, 103149. https://doi.org/10.1016/j.apergo. 2020.103149

[46]   Mesko, B., Drobni, Z., Bényei, E., Gergely, B., and Györffy, Z. (2017) Digital health is a cultural transformation of traditional healthcare. *mHealth* 3, 38. https://doi.org/10.21037/mhealth.2017.08.07

[47]   Mohanty, A., and Mishra, S. (2022) A comprehensive study of explainable artificial intelligence in healthcare. In *Augmented Intelligence in Healthcare: A Pragmatic and Integrated Analysis.* Springer Nature Singapore, 475–502. https://doi.org/10.1007/978-981-19-1076-025

[48]   Yang, G., Ye, Q., and Xia, J. (2022) Unbox the black-box for the medical explainable AI via multi-modal and multi-centre data fusion: A mini-review, two showcases and beyond. *Information Fusion* 77, 29–52. https://doi.org/10.1016/j.inffus.2021.07.016

[49]   Sagi, O. and Rokach, L. (2020) Explainable decision forest: Transforming a decision forest into an interpretable tree. *Information Fusion* 61, 124–138. https://doi.org/10.1016/j.inffus.2020.03.013

[50]   Jovanovic, M., Radovanovic, S., Vukicevic, M., Poucke, S. V., and Delibasic, B. (2016) Building interpretable predictive models for pediatric hospital readmission using tree-lasso logistic regression. *Artificial Intelligence in Medicine* 72, 12–21. https://doi.org/10.1016/j.artmed. 2016.07.003

[51]   Shouval, R., Hadanny, A., Shlomo, N., Iakobishvili, Z., Unger, R., Zahger, D., Alcalai, R., Atar, S., Gottlieb, S., Matetzky, S., Goldenberg, I., and Beigel, R. (2017) Machine learning for prediction of 30-day mortality after ST elevation myocardial infraction: An acute coronary syndrome Israeli survey data mining study. *International Journal of Cardiology* 246, 7–13. https://doi.org/10.1016/j.ijcard.2017.05.06

[52]   Lundberg, S. M., and Lee, S. -I. (2017) A unified approach to interpreting model predictions. In *Proceedings of the 31st International Conference on Neural Information Processing Systems (NIPS'17).* Curran

Associates Inc., Red Hook, NY, USA, 4768–4777. https://dl.acm.org/doi/epdf/10.5555/3295222.3295230

[53] Shapley, L. S. (1953) A value for n-person games. *Contributions to the Theory of Games (Am-28)* II, 307–318. https://doi.org/10.1515/ 9781400881970-018

[54] Linardatos, P., Papastefanopoulos, V., and Kotsiantis, S. (2020) Explainable AI: A review of machine learning interpretability methods. *Entropy* 23(1), 18. https://doi.org/10.3390/e23010018

[55] Raza, A., Tran, K. P., Koehl, L., and Li, S. (2022) Designing ECG monitoring healthcare system with federated transfer learning and explainable AI. *Knowledge-Based Systems* 236, 107763. https://doi.org/10.1016/j.knosys.2021.1

[56] Khodabandehloo, E., Riboni, D., and Alimohammadi, A. (2021) HealthXAI: Collaborative and explainable AI for supporting early diagnosis of cognitive decline. *Future Generation Computer Systems* 116, 168–189. https://doi.org/10.1016/j.future.2020.10.030

[57] Iwasawa, Y., Nakayama, K., Yairi, I., and Matsuo, Y. (2017) Privacy issues regarding the application of DNNs to activity-recognition using wearables and its countermeasures by use of adversarial training. In *Proceedings of the 26th International Joint Conference On Artificial Intelligence.* International Joint Conferences on Artificial Intelligence Organization. https://www.ijcai.org/Proceedings/2017/0268.pdf

[58] Chen, K., Zhang, D., Yao, L., Guo, B., Yu, Z., and Liu, Y. (2021) Deep learning for sensor-based human activity recognition. *ACM Computing Surveys* 54(4), 1–40.

[59] Zhang, D., Yao, L., Chen, K., Long, G., and Wang, S. (2019) Collective protection: Preventing sensitive inferences via integrative transformation. In *2019 IEEE International Conference on Data Mining (ICDM).* IEEE. https://doi.org/10.1109/icdm.2019.00197

[60] Gati, N. J., Yang, L. T., Feng, J., Nie, X., Ren, Z., and Tarus, S. K. (2021) Differentially private data fusion and deep learning framework for cyber–physical–social systems: State-of-the-art and perspectives. *Information Fusion* 76, 298–314. https://doi.org/10.1016/j.inffus.2021. 04.017

[61] Hossein, K. M., Esmaeili, M., Dargahi, T., and Khonsari, A. (2019) Blockchain-based privacy-preserving healthcare architecture. In *2019 IEEE Canadian Conference of Electrical and Computer Engineering (CCECE).* IEEE, 2576–7046. https://doi.org/10.1109/ CCECE.2019.8861857

[62] Ul Hassan, M., Rehmani, M. H., and Chen, J. (2020) Differential privacy in blockchain technology: A futuristic approach. *Journal of Parallel and Distributed Computing* 145, 50–74. https://doi.org/10.1016/j.jpdc.2020.06.003

[63]   Singh, S., Rathore, S., Alfarraj, O., Tolba, A., and Yoon, B. (2021) A frame-work for privacy-preservation of IoT healthcare data using federated learning and blockchain technology. *Future Generation Computer Systems* 129, 380–388. https://doi.org/10.1016/j.future.2021.11.028

[64]   Gawlikowski, J., Tassi, C. R. N., Ali, M., Lee, J., Humt, M., Feng, J., Kruspe, A., Triebel, R., Jung, P., Roscher, R., and Shahzad, M. (2021) *A Survey of Uncertainty in Deep Neural Networks.* https://doi.org/10.48550/ARXIV.2107.03342

[65]   Hüllermeier, E., and Waegeman, W. (2021) Aleatoric and epistemic uncertainty in machine learning: An introduction to concepts and meth-ods. *Machine Learning* 110(3), 457–506. https://doi.org/10.1007/s10994-021-05946-3

[66]   Abdar, M., Pourpanah, F., Hussain, S., Rezazadegan, D., Liu, L., Ghavamzadeh, M., Fieguth, P., Cao, X., Khosravi, A., Acharya, U. R., Makarenkov, V., and Nahavandi, S. (2021) A review of uncertainty quan-tification in deep learning: Techniques, applications and challenges. *Information Fusion* 76, 243–297. https://doi.org/10.1016/j.inffus.2

[67]   Wang, H., and Yeung, D.-Y. (2016) *A Survey on Bayesian Deep Learning.* https://arxiv.org/abs/1604.01662

[68]   Wang, Y., and Zheng, Y. (2018) Modeling RFID signal reflection for contact-free activity recognition. *Proceedings of the ACM on Interactive, Mobile, Wearable and Ubiquitous Technologies* 2(4), 1–22. https://doi.org/10.1145/3287071

[69]   Begoli, E., Bhattacharya, T., and Kusnezov, D. (2019) The need for uncertainty quantification in machine-assisted medical decision making. *Nature Machine Intelligence* 1(1), 20–23. https://doi.org/10.1038/s42256-018-0004-1

Remote Patient Monitoring by Artificial Intelligence: Bringing a Paradigm Shift to Healthcare

Abstract

Every sector is undergoing a revolution due to technological advancements, aiming to provide superior products and services to consumers. The healthcare sector is one such sector, with contemporary technology being used to revolutionize traditional approaches to the care and treatment of patients. The modern healthcare system's overarching goal has expanded to include more than just the treatment of individual diseases. Sensor-enabled systems, integrating artificial intelligence (AI) into Internet-of-Things (IoT)-based smart healthcare systems, creates the orchestration of healthcare services and allows healthcare professionals to monitor patients' daily activities in real-time, thereby allowing for faster responses to patients' healthcare needs. In this chapter, we examine what it takes to build effective AI-based services for the healthcare industry and provide a narrative assessment of healthcare services that includes AI-based services in their operations. AI is showing its worth in the healthcare industry by enhancing healthcare outcomes, aiding caregivers in their job, and decreasing healthcare expenses. The market potential for AI in healthcare is likewise substantial, with a projected compounded annual growth rate of 28% worldwide. This chapter will compile findings from a variety of healthcare sectors, such as financial, health improvement, and care results, and then provide solutions and highlight critical success elements for incorporating AI techniques into medical practice. This research demonstrates how using AI in healthcare might improve outcomes while decreasing overall healthcare costs.

2.1 Introduction

Healthcare technology has advanced rapidly. It can handle major changes. Technology can be utilized to provide more reliable, efficient, and effective treatments for infectious diseases, cancer, radiography, and risk evaluation. Heart disease, diabetes, and Alzheimer's will cause 66% of fatalities by 2030. Telehealth—including virtual healthcare—improves communication between doctors, clinics, and patients. Clinics, physicians, and patients can interact, track, and follow up on care plans using electronic communication technologies, maximizing virtual involvement in medical treatment. Technology can now be employed for therapy, pre-operative planning, and remote monitoring, according to the WHO [1]. Regularly monitoring chronic disease patients prevents life-threatening scenarios. The Internet of Things, which connects everyday objects to the globe, emerged this decade, and is used in remote health monitoring [2, 3], parking management [4], intelligent homes [5], cities [6], environments [7], workplaces [8], and farms [9], among others. IoT can monitor environmental and physiological aspects to manage healthcare. Real-time applications are using IoT systems because of their simplicity. Real-time data analysis is possible with the IoT. Sensored, internet-connected "objects" do this [10, 11]. Figure 2.1 shows IoT real-time systems.

Figure 2.1: IoT applications in daily life.

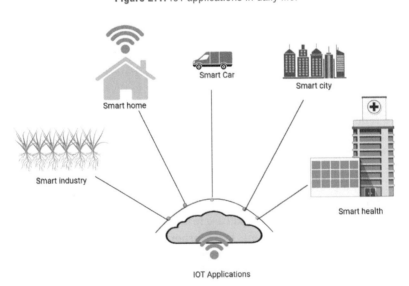

"Health is wealth" still applies today. A hurried lifestyle, growing pollution, and epidemic and pandemic diseases have caused poor life quality. Over 90% of the population has been exposed to contamination [12]. The mechanical revolution in agriculture and population growth have impoverished most people. Thus, healthy lifestyle monitoring, improvement, and promotion are needed. Industry 5.0 and 5G communications technology have enabled affordable observational sensors and real-time data collecting [13]. Figure 2.2 shows SHM applications.

Figure 2.2: Benefits and applications of SHM.

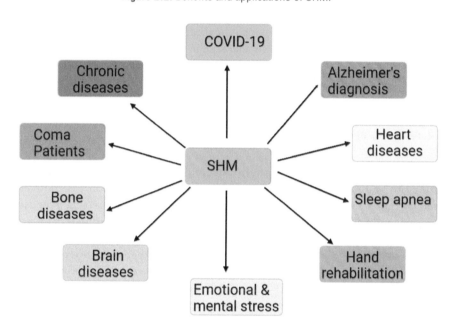

The internet of medical things (IoMT) powers smart health monitoring in the IoT [14]. The Internet of Things was inspired by connecting gadgets to a network for data exchange. Networked healthcare devices form the "World Wide Web" of healthcare devices. Telemedicine, remote medical diagnosis, medical nursing, physical rehabilitation, and patient monitoring are examples. IoT and SHM enhancements to hospital medical systems allow patients to be treated more accurately and carefully than before. Our technologies have prioritized life safety and cheap inspection costs for decades. IoT, or the Internet of Things, could lower costs for a variety of cutting-edge techniques, such as SHM and general smart monitoring (GSM), as well as methods that are similar, such as using geophysical methods like electrical resistivity, ultrasonic surface waves, and ground penetrating radar to examine subsurface structures.

The "Internet of Things" (IoT) is a network of electronic and wireless devices that may gather, store, and analyze personal data for diagnosis and treatment [15]. Wireless sensors and gadgets capture patient data. SHM is a medical breakthrough that uses IoT accelerometers, sensors, blinking eye monitors, temperature monitors, etc. to monitor patients' health, particularly coma patients. This technique may help the elderly and chronically ill [16]. The elderly and chronically ill have dramatically increased their need for remote health monitoring systems in recent decades. Hospitalizations have increased as more individuals worldwide seek medical care. According to a report, 770,000 Americans die each year, or a daily increase in mortality [6]. This is brought on by improper medicine use, delayed treatment, and wrong dosage, among other things. SHM models reduce staff and medical professional's workloads [18-20].

2.2 Attributes of SHM and IoMT

Smart health monitoring devices help clinicians remotely monitor patients' conditions and reduce unnecessary hospital visits. Hackers cannot access data created by these devices. These innovations address healthcare's rising expenses, a major medical advancement. Due to rising expectations and beliefs in technology and digitization, IoMTs and SHM will continue to contribute to growth and development. SHM is cutting-edge technology that might remotely run hospitals to save patients with life-threatening diseases like heart attacks, asthma attacks, diabetes, and more. IoMT and SHM data must be regressed and computed to get multiple results. Data analysis and calculation can avoid disease spread, chronic disease, patient deaths, and future health monitoring. Health care data analysis requires deep learning and other AI [21]. The SHM network has many frameworks and suggested architects for healthcare data analysis [22–24].

2.3 The IoTM and SHM Frameworks

Disassembled gadgets may include:
1. Sharing medical records and other data is its foundation, based on the transmission of health-related information or data from phone to phone, or from patients to doctors.
2. Publisher

Figure 2.3: Components of IoT based networks.

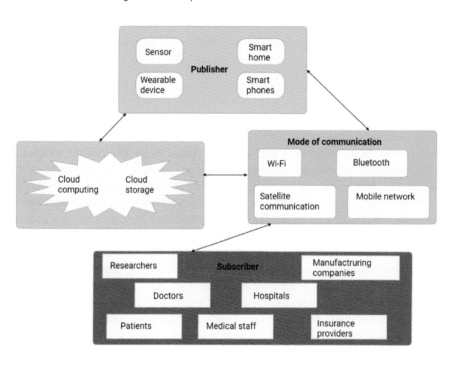

3.   Broker
4.   Subscriber.

Sensors and patient records include BP, ecological acuity, glucose, and more. On this arrangement, the broker receives data from the publisher and sometimes stores it in the cloud. The last type is a subscriber, with which smartphones, PCs, tablets, and wearable gadgets may identify and utilize to continuously monitor data from a source. Figure 2.3 displays IoMT classification.

A remote health monitoring system (RHMS) can send and receive data, making it suitable for use in private homes and medical facilities. It can monitor a wide range of symptoms. The MHMS uses cellphones, laptops, and pocket PCs as its primary processing unit. Wearable technology like digital watches, wrist-bands, eye blinkers, oximeters, pulse trackers, and others collects meaningful and original health data, using real-time data. These are recent developments in medical technology. In today's world, GHMS plays a crucial function in keeping track of common ailments. As a result, fewer people will visit hospitals for common ailments. Today's health-conscious society allows consumers to use online medical services and other tools to solve their health issues. They can also use one of several free medical applications for all major operating systems to search nearby hospitals and doctors. Medical IoTs require specific parts:

1. Aquisition of data
2. Channel of interaction
3. Server.

The Internet of Things (IoT) has three layers for proper data collection. Body sensors and wearable technology are included in the device layer. The data is semi-processed in the fog layer and stored in the cloud layer, which has high computing resources. The fog layer's semi-processed data is turned into pertinent information and numerous intended goals are achieved by employing a lot of processing power. An extremely cost-effective solution for medical services, the cloud also functions as an on-demand service. Data acquisition is the first step in gathering all patient data utilizing different types of smart health devices, which essentially fall into two categories: wireless sensors and wearables, including smart phones, smart cameras, microphones, and raspberry Pis with cameras being employed as wireless sensors. These sensors are especially useful for patients with chronic illnesses, those in comas, and newborns for smart care, for smart homes for independent elders, for measuring respiratory rate and cardiac rate by using air matter sensors, toilet seats integrated with an ECG monitor, cardiac, beat monitor, respiratory monitor, and many more. Wearable technology, including digital watches, clothing, fitness tracker bands, etc., falls within the WHMS and GHMS categories and connects to other devices to capture a wide range of data. In some circumstances, a smart vest is used to monitor physiological data such as an electrocardiogram (ECG) without the use of skin-contacting gel, blood pressure, body temperature, and galvanic skin response. It also assists in determining cardiac output using eco-doppler to analyse and improve the performance of athletes and people with disabilities.

SHM and IoTs may aid long-term ailments like Parkinson's, Alzheimer's, and dementia. Dementia, the most common chronic illness, affects 25 million people globally. Dementia symptoms include memory loss, difficulty making decisions and solving issues, melancholy, and anxiety. Many illnesses cause dementia; Parkinson's and Alzheimer's are common causes. Dementia has numerous diseases as causes, many of which are unknown. The two most prevalent causes are Alzheimer's and Parkinson's illnesses. The SHM plays a crucial role in these situations, which is an outstanding accomplishment in the medical field [25]. IoMT data could be misused for personal gain and researchers struggle with SHM data security. Figure 2.4 categorizes SHM.

Figure 2.4: Subclasses of smart health monitoring (SHM).

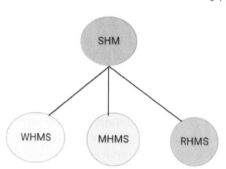

2.4 Data Security in SHM

IoMT and SHM networks generate big data or healthcare data, which requires lots of computing power and storage. Cloud computing and storage solve cloud healthcare data management, but most of this information is hidden from patients. SHM concerns are data security and privacy. Preventing users, people, and organizations from exploiting sensitive data for personal gain is best. Data security includes physical, authentication, network, computer, and secure storage. Encoding, decoding, genetic algorithms, encrypted data storage, and cryptography have increased in use. Trust is a constant issue, hence the majority of security and privacy frameworks offer third parties. Block chain and the interplanetary file system for banking and finance data sharing have grown in popularity [26–29]. Block chains are chains of blocks of data and blocks include transaction records. Cryptographic methods check new blocks as they are added to the chain, which includes the whole public record of transaction history, and are time-stamped. Because each block contains the hash value of the previous block and the current block, block chain data transport is secure and dependable. Block chain technology is successfully employed in manufacturing, management, logistics, and healthcare, and many medical care services and SHM network models have been tested and are publicly available [30–32].

2.5 Remote Heart Rate Tracking

Recent technical advances allow online data collecting from heart disease patients and at-risk individuals. Symptoms, body mass index (BMI), physical activity (PA), pulse rate (HR), HR regularity (HRV), heart sounds (HRS),

respiratory rate (RR), RR interval, electrocardiogram (ECG), oxygen saturation (SpO2), and sleep quality may all be affected. These technologies also
incorporate an expanding range of wearables, like a watch or a patch [33,34].
Point-of-care diagnostics have helped treat patients outside hospitals. Cardiovascular implanted electronic devices can provide nearly constant streams
of physiological data to patients with a history of heart failure or a cardiac
arrhythmia near-death experience (such as use of a defibrillator). Traditional
or machine learning may help detect clinically significant occurrences. Remote
monitoring data may accelerate preventative clinical interventions like patient
adherence, drug treatment adjustments, and early clinical face-to-face reviews.
Direct-to-home medication delivery services like Amazon's PillPack complete
the monitoring-to-therapy cycle. Smartphone apps are helping more patients to
manage their health, and self-management without health-care team involvement may become the norm as more people use digital health services. Health
policymakers are hopeful that this customized strategy will improve population
health and save money.

2.6 Sensor Enabled IoT-based Devices

There are many connected health monitors that upload data to the internet.
The smartphone app lets users measure fitness, number of steps, distance run
or cycled, and sleep [35]. Also users of certain high-variant smart devices will
receive notifications about female health status, stress conditions, and guided
meditation during stressful conditions [36], fitness tracking, and evening exer-
cise based on the day's number of calories burnt, as well as continuous heart
rate and blood pressure monitoring, including ECG in higher variants with fall
detection alarms [36], when there is a drastic shift in vitals, as evidenced by
the appearance of a new, different, or abnormal pattern. These smart devices
monitor heart, blood pressure, activity, sleep, and female health, as well as
oxygen saturation levels [36]. Technology can treat metabolic disorders like
diabetes. To monitor their condition, diabetics may utilize daily finger-prick
blood sugar testing with new strips. The procedure is painful and expensive.
Innovative smart technological solutions may solve the problem. Smart glu-
cometers eliminate finger pricking, sterile needles, and strips by storing past
readings [38].

2.7 Superior Medical Technology

"Smart healthcare" (also known as "intelligent healthcare") "uses new,
sophisticated, and advanced technology to remotely access knowledge, link

individuals, resources, and organizations relevant to healthcare" [39]. Smart healthcare uses cutting-edge technologies including wearables, the Internet of Things, and mobile Internet. Smart healthcare uses cutting-edge technology to "connect to patients," "manage availability," and "improve medical operations." The "leapfrog solution" for large-data remote patient monitoring is portable, low-cost ICT equipment such laptops, PCs, and mobile phones [40]. This has raised awareness and support for healthcare wearable devices and patient monitoring systems based on smartphone apps that connected doctors, physicists, and smart wearable health care equipment. Including hospitals on a single network for real-time access [41]. In cardiac monitoring smart systems, which monitor a patient's pulse, blood oxygen levels, and electrocardiogram (ECG) [36,37], a cloud-based, Internet-of-Things-enabled wearable device constantly generates data, which is stored. An artificial intelligence algorithm uses supervised learning to uncover previously unseen patterns in the data, turning the system into a smart one that can accurately predict fails.

2.8 Smart Health Monitoring

IoT-based smart devices with AI are a danger to healthcare delivery and administration because they use compatible sensors to predict future problems. For arrhythmia, artificial intelligence systems track heart activity and use supervised learning to identify trends. When the pattern changes, it will warn users to a possible cardiac rhythm abnormality and recommend a doctor visit [37]. When the AI algorithm detects a sudden, significant change in the user's heart rate and breathing pattern, the Samsung Active Watch 2 displays stress indicators and recommends guided meditation via the Samsung Health mobile app [36]. Sensor data and machine learning help health monitoring too. Wearable healthcare solutions with smart sensors and AI will help doctors make quick but accurate treatment decisions for in-patients by documenting users' everyday activities and physiological symptoms. Thus, AI and analytics on vast volumes of lab data and IPD papers utilizing supervised learning from prior patients may minimize hospital stays and re-admission rates [42].

2.9 Service Coordination in Healthcare

Industry 4.0 sensors are affecting many industries, including healthcare. Modern healthcare requires hospital consultations, admissions, critical care, ambulatory care, laboratory testing, CT scans, radiology tests, and radiation therapy for acute and chronic illnesses. Hospitals will become minor elements of a

Table 2.1: The significance of smart healthcare systems.

	Application to a healthcare setting	Impact
Remote consultation	Includes eHealth, mHealth Services	Low
	Remote patient—physician consultation over any audio, video aid	Low
IoT enabled patient monitoring	Includes patient monitoring with aid of wearable sensors	Moderate
	Noninvasive digital technologies for home care	Moderate
	AI enabled algorithms which predicts the fall detection in the case of metabolic disorders	High
Orchestrated services technology	Inter-connected medical staff to provide 360 degree services	High
	AI-enabled algorithms which send notification to care professionals about an emergency situation at the patients' end	High

large healthcare ecosystem [43]. Healthcare silos can be connected. Consultation, laboratory testing, imaging, and ambulatory care patients will move, but major surgery and critical care patients will stay. Work that is disconnected requires coordination. People are drawn to care service orchestration by real-time digital technologies [44,45]. An interactive stakeholders method has been developed for growth and coordinated information and service flow. Integrative care service orchestration enables healthcare workers and administrators to see the whole system, linking distributed systems to a standardized network. By connecting disparate systems to a common network, integrative care service orchestration enables healthcare professionals and administrators to see the entire system and visual display. Networked facilities benefit the network. Care service orchestration requires upstream and downstream cooperation, where doctors can communicate quickly [46]. Healthcare organizations must fully integrate doctors and nurses. Good care requires knowing current therapeutic developments, procedures, hospital capacity, diagnostic tests (pathology, radiology, etc.), and patient feedback [47]. Table 2.1 ranks smart systems technology for healthcare use.

2.10 Conclusion

Our extensive analysis found that AI may save healthcare costs, provide preventative care, reduce healthcare worker workload, and provide more accurate, faster diagnoses. Healthcare costs are growing, requiring AI services. The aging population in industrialized nations will increase chronic diseases that require expensive treatment. The poor, aged, and most emerging nations lack modern, effective healthcare. AI in healthcare research and IT can save money and improve health and quality of life. Every researched healthcare service needs AI advances. This study predicted AI-enabled intelligent healthcare services. A smart healthcare system is being developed using AI and IOT-based health services with sensors. We have laid down the groundwork for future healthcare services by analyzing AI, ML, smart systems, and IoT research. Sensor-enabled, Internet-of-Things-based wearables and artificial intelligence allow a network of medical specialists to remotely monitor a patient, resulting in synchronized healthcare services.

References

[1] Perumal, K. and Manohar, M. (2017) A survey on internet of things: case studies, applications, and future directions. In *Internet of Things: Novel Advances and Envisioned Applications* . Cham: Springer, 281-297.

[2] Rahaman, A., Islam, M.M., Islam, M.R., Sadi, M.S. and Nooruddin, S. (2019) Developing IoT based smart health monitoring systems: A review. *Rev. d'Intelligence Artif.* 33(6), 435-440.

[3] Islam, S.R., Kabir, Md. H., Hossain, Md., and Kwak, D. (2015) The Internet of Things for health care: A comprehensive survey. *IEEE Access* 3, 678-708.

[4] Lin, T., Rivano, H. and Le Mouël, F. (2017) A survey of smart parking solutions. *IEEE Transactions on Intelligent Transportation Systems* 18(12), 3229-3253.

[5] Zualkernan, I.A., Rashid, M., Gupta, R. and Alikarar, M. (2017) A smart home energy management system using IoT and big data analytics approach. *IEEE Transactions on Consumer Electronics* 63, 426-434.

[6] Zanella, A., Bui, N., Castellani, A., Vangelista, L. and Zorzi, M. (2014) Internet of things for smart cities. *IEEE Internet Things J.* 1(1), 22–32.

[7] Mois, G., Folea, S. and Sanislav, T. (2017) Analysis of three IoT-based wireless sensors for environmental monitoring. *IEEE Transactions on Instrumentation and Measurement* 66(8), 2056-2064.

[8] Chen, B., Wan, J., Shu, L., Li, P., Mukherjee, M. and Yin, B. (2017) Smart factory of industry 4.0: Key technologies, application case, and challenges. *IEEE Access* 6, 6505-6519.

[9] Ayaz, M., Ammad-Uddin, M., Sharif, Z., Mansour, A. and Aggoune, E.H.M. (2019) Internet-of-Things (IoT)-based smart agriculture: Toward making the fields talk. *IEEE Access* 7, 129551-129583.

[10] Hasan, M., Islam, M.M., Zarif, M.I.I. and Hashem, M.M.A. (2019) Attack and anomaly detection in IoT sensors in IoT sites using machine learning approaches. *Internet of Things* 7, 100059.

[11]   Nooruddin, S., Islam, M.M. and Sharna, F.A. (2020) An IoT based device-type invariant fall detection system. *Internet of Things* 9, 100130.

[12]   Dhimal, M., Chirico, F., Bista, B., Sharma, S., Chalise, B., Dhimal, M.L., Ilesanmi, O.S., Trucillo, P. and Sofia, D. (2021) Impact of air pollution on global burden of disease in 2019. *Processes* 9(10), 1719.

[13]   Tabaa, M., Monteiro, F., Bensag, H. and Dandache, A. (2020) Green Industrial Internet of Things from a smart industry perspectives. *Energy Reports* 6, 430-446.

[14]   De Michele, R. and Furini, M. (2019) IoT healthcare: Benefits, issues and challenges. In *Proceedings of the 5th EAI International Conference on Smart Objects and Technologies for Social Good* 160-164.

[15]   Haricha, K., Khiat, A., Issaoui, Y., Bahnasse, A. and Ouajji, H. (2020) Towards smart manufacturing: Implementation and benefits. *Procedia Computer Science* 177, 639-644.

[16]   Portnoy, J.M., Pandya, A., Waller, M. and Elliott, T. (2020. Telemedicine and emerging technologies for health care in allergy/immunology. *Journal of Allergy and Clinical Immunology* 145(2), 445-454.

[17]   Baig, M.M. and Gholamhosseini, H. (2013) Smart health monitoring systems: an overview of design and modeling. *Journal of Medical Systems* 37(2), 1-14.

[18]   Rahaman, A., Islam, M.M., Islam, M.R., Sadi, M.S. and Nooruddin, S. (2019) Developing IoT Based Smart Health Monitoring Systems: A Review. *Rev. d'Intelligence Artif.* 33(6), 435-440.

[19]   Albahri, A.S., Albahri, O.S., Zaidan, A.A., Zaidan, B.B., Hashim, M., Alsalem, M.A., Mohsin, A.H., Mohammed, K.I., Alamoodi, A.H., Enaizan, O. and Nidhal, S. (2019) Based multiple heterogeneous wearable sensors: A smart real-time health monitoring structured for hospitals distributor. *IEEE Access* 7, 37269-37323.

[20]   Hassanalieragh, M., Page, A., Soyata, T., Sharma, G., Aktas, M., Mateos, G., Kantarci, B. and Andreescu, S. (2015, June. Health monitoring and management using Internet-of-Things (IoT) sensing with cloud-based processing: Opportunities and challenges. In *2015 IEEE International Conference on Services Computing* IEEE, 285-292.

[21]   Azimi, M., Eslamlou, A.D. and Pekcan, G. (2020) Data-driven structural health monitoring and damage detection through deep learning: State-of-the-art review. *Sensors* 20(10), 2778.

[22]   Jang, H.J. and Cho, K.O. (2019) Applications of deep learning for the analysis of medical data. *Archives of Pharmacal Research* 42(6), 492-504.

[23]   Yue, L., Tian, D., Chen, W., Han, X. and Yin, M. (2020) Deep learning for heterogeneous medical data analysis. *World Wide Web* 23(5), 2715-2737.

[24]   Ravì, D., Wong, C., Deligianni, F., Berthelot, M., Andreu-Perez, J., Lo, B. and Yang, G.Z. (2016) Deep learning for health informatics. *IEEE Journal of Biomedical and Health Informatics* 21(1), 4-21.

[25]   Ravì, D., Wong, C., Deligianni, F., Berthelot, M., Andreu-Perez, J., Lo, B. and Yang, G.Z. (2016) Deep learning for health informatics. *IEEE Journal of Biomedical and Health Informatics* 21(1), 4-21.

[26]   Wang, L., Sha, L., Lakin, J.R., Bynum, J., Bates, D.W., Hong, P. and Zhou, L. (2019) Development and validation of a deep learning algorithm for mortality prediction in selecting patients with dementia for earlier palliative care interventions. *JAMA Network Open* 2(7), e196972-e196972.

[27]   Ratta, P., Kaur, A., Sharma, S., Shabaz, M. and Dhiman, G. (2021) Application of blockchain and internet of things in healthcare and medical sector: applications, challenges, and future perspectives. *Journal of Food Quality* 2021.

[28]   Bhola, J., Soni, S. and Cheema, G.K. (2019) Recent trends for security applications in wireless sensor networks–a technical review. In *2019 6th International Conference on Computing for Sustainable Global Development (INDIACom)* IEEE, 707-712.

[29]   Bhola, J., Soni, S. and Cheema, G.K. (2020) Genetic algorithm based optimized leach protocol for energy efficient wireless sensor networks. *Journal of Ambient Intelligence and Humanized Computing* 11(3), 1281-1288.

[30]   Kumar, A., Jagota, V., Shawl, R.Q., Sharma, V., Sargam, K., Shabaz, M., Khan, M.T., Rabani, B. and Gandhi, S. (2021) Wire EDM process parameter optimization for D2 steel. *Materials Today: Proceedings* 37, 2478-2482.

[31]   Pham, H.L., Tran, T.H. and Nakashima, Y. (2018) A secure remote healthcare system for hospital using blockchain smart contract. In *2018 IEEE Globecom Workshops (GC Wkshps)* IEEE, 1-6.

[32]   Griggs, K.N., Ossipova, O., Kohlios, C.P., Baccarini, A.N., Howson, E.A. and Hayajneh, T. (2018) Healthcare blockchain system using smart contracts for secure automated remote patient monitoring. *Journal of Medical Systems* 42(7), 1-7.

[33]   Antwi, M., Adnane, A., Ahmad, F., Hussain, R., ur Rehman, M.H. and Kerrache, C.A. (2021) The case of hyperledger fabric as a blockchain solution for healthcare applications. *Blockchain: Research and Applications* 2(1), 100012.

[34]   Mueller, B. (2020) Telemedicine arrives in the UK: '10 years of change in one week'. *The New York Times* https://www.nytimes.com/2020/04/04/world/europe/telemedicine-uk-coronavirus.html.

[35]   Krittanawong, C., Rogers, A.J., Johnson, K.W., Wang, Z., Turakhia, M.P., Halperin, J.L. and Narayan, S.M. (2021) Integration of novel monitoring

devices with machine learning technology for scalable cardiovascular management. *Nature Reviews Cardiology* 18(2), 75-91.

[36] Deshkar, S., Thanseeh, R.A. and Menon, V.G. (2017) A review on IoT based m-Health systems for diabetes. *International Journal of Computer Science and Telecommunications* 8(1), 13-18.

[37] Bhatt, V. and Chakraborty, S. (2021) Real-time healthcare monitoring using smart systems: A step towards healthcare service orchestration Smart systems for futuristic healthcare. In *2021 International Conference on Artificial Intelligence and Smart Systems (ICAIS)* IEEE, 772-777.

[38] Apple (2019) ECG app and irregular heart rhythm notification available today on Apple Watch. *Apple Newsroom.* https://www.apple.com/uk/newsroom/2019/03/ecg-app-and-irregular-rhythm-notification-on-apple-watch-available-today-across-europe-and-hong-kong/

[39] Kukkar, D., Zhang, D., Jeon, B.H. and Kim, K.H. (2022) Recent advances in wearable biosensors for non-invasive monitoring of specific metabolites and electrolytes associated with chronic kidney disease: Performance evaluation and future challenges. *TrAC Trends in Analytical Chemistry* 116570.

[40] Tian, S., Yang, W., Le Grange, J.M., Wang, P., Huang, W. and Ye, Z. (2019) Smart healthcare: making medical care more intelligent. *Global Health Journal* 3(3), 62-65.

[41] Yeong, Y.C., Kalid, K.S. and Sugathan, S.K. (2019) Cryptocurrency acceptance: A case of Malaysia. *International Journal of Engineering and Advanced Technology* 8(5), 28-38.

[42] Bhatt, V. and Chakraborty, S. (2020) Importance of trust in IoT based wearable device adoption by patient: an empirical investigation. In *2020 Fourth International Conference on I-SMAC (IoT in Social, Mobile, Analytics and Cloud)(I-SMAC)* IEEE, 1226-1231.

[43] Wehbe, Y., Al Zaabi, M. and Svetinovic, D. (2018, November. Blockchain AI framework for healthcare records management: constrained goal model. In *2018 26th Telecommunications Forum (TELFOR)* IEEE, 420-425.

[44] Friesdorf, M., Deetjen, U., Sawant, A., Gilbert, G. and Niedermann, F. (2019) Digital health ecosystems: A payer perspective. *McKinsey & Company* 2 August.

[45] Sezen, B. (2008) Relative effects of design, integration and information sharing on supply chain performance. *Supply Chain Management: An International Journal* 13(3).

[46] Prajogo, D. (2012) Olhager J, Supply chain integration and performance: The effect of long-term relationships, IT and sharing, and logistics integration. *International Journal of Production Economies* 135(1), 514-522.

CHAPTER

3

Advancement in AI Enables Hospital Movement at Home

Abstract

This chapter explores the possibilities of health "smart" homes (HSH) via telemedicine and other forms of modern communication. Persons requiring special facilities who are committed to maintaining their independence may get medical support from HSH systems. Home-based patients' diverse requirements necessitate specialized technologies. These demands need a distributed strategy and several hardware and software approaches. We briefly discuss home health care's innovative information, communication, and data-acquisition systems. Specifically, we discuss in detail the technical, monetary, and social requirements of HSH. In addition to a brief summary of our own smart home and telehealthcare information system project, examples of HSH initiatives are provided. The impact of COVID-19 has been disruptive, forcing people, organizations, and governments to make significant adjustments to their previous ways of doing things. These shifts have proven to be a driver of technological and innovative progress. In this chapter, researchers examine how the pandemic may impact the growth of the Internet of Things across a variety of settings, such as healthcare, smart homes, smart buildings, and smart cities.

3.1 Introduction

Recent debates have focused on artificial intelligence (AI) and its effects in several industries. However, Ramon Llull's idea of a thinking machine in 1300

CE and Aristotle's syllogisms in 300 BCE predate artificial intelligence. In recent years, AI software development has increased despite a fall in interest and funding in the 1970s and 1980s. China has invested billions in AI business centers. Since they have sophisticated automated systems, smart houses can monitor and adjust household activities for convenience, comfort, and energy savings [1]. Smart home technologies evaluate household data, inform users, and expand home system control [2]. AI is a machine's ability to comprehend and adapt to its surroundings to increase its chances of success [3]. AI should reason, behave logically, and think like a human [3]. Smart home products use all technology, and their AI functions include activity detection, data processing, voice recognition, image identification, decision making, and prediction. Smart home devices that use artificial intelligence (AI) might be able to detect human activity. When it notices anything unusual, like human activity, it examines sensor data to determine the cause. Devices such as Hive Link and Essence CareHome analyze sensor data to locate abnormalities like human activity and recognize the activity. Data processing AI relies on data analysis, data collection, and underlying linkages. AI's voice-driven speech recognition lets consumers talk to it, for instance to ask about the weather, order online, or call a taxi. Josh Micro, Ivee Sleek, Jibo, Athom Homey, Amazon Alexa, Google Home, and others use voice recognition. Image identification uses AI for face recognition, emotion detection, biometrics, and scene understanding. Human behavior and body structure are measured and analyzed. Lighthouse, Google Cam, Honeywell Smart Home Security System, Tend Secure Lynx Indoor Camera, Canary All-in-One, and Netatmo Welcome Indoor Security Camera use it. AI makes decisions, and it can process data [4]. A smart security system can alert the owner via smartphone or call the police if a house camera detects an intruder. These systems must react quickly and efficiently. Arlo Ultra, Ecobee4, and VELUX roof windows and shades use it. Sensors placed in homes provide assistance with forecasting by gathering data on individuals' daily routines and habits. A computer network records environmental data in a database for an intelligent agent to evaluate it for patterns, forecasts, and trends. This data can assist smart home software choose and automate tasks [5]. Smart home products and services are largely for climate control. There have been attempts to create the most advanced smart house [6]. Telemedicine's multiple uses in telepresence, teleconsulting and diagnostics, remote monitoring, real-time medical scanning, telesurgery, and remote primary care explain its rapid expansion. These subfields use innovative technology to improve health. These systems must balance medical needs with technology improvements. System redesigns should prioritize patients over technology [7].

Figure 3.1: Core elements of a smart city.

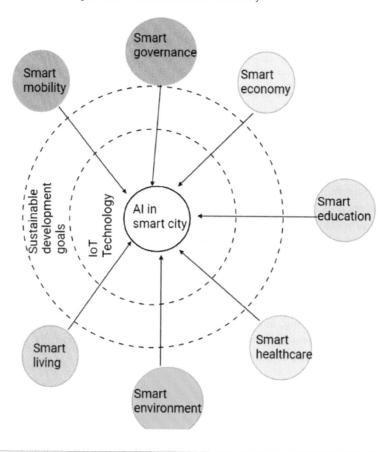

Figure 3.1 shows smart city AI uses. Since 2015, healthcare research has decreased but intelligent interaction research has increased. Energy management research is expanding. Future smart houses may focus more on how people and the environment interact and how to make structures more customized and sustainable.

3.2 "Health Smart Home"

A health smart home (HSH) has electronics to operate home medical automation instruments remotely. A home with automated medical gadgets and sensors would monitor the patient's health and safety. It also has a remote control center (RCC) and a local intelligence unit (LIU) that searches sensor data for crucial or suspicious events. Doctors, pharmacists, emergency specialists, mental health workers, family or volunteer caregivers, technicians, and system administrators can call the RCC in an emergency (Figure 3.2). The RCC can take many different forms, depending on the local health policy, including a specialized emergency unit in a community hospital, installations and facilities run by a private healthcare provider, or even a particular service in a top university hospital (like the hospital at home services in France). The RCC provides tech support for hardware and system maintenance and live human support for emergency calls to improve patient-medical staff contact [8-10]. The HSH provides several medical and social services in addition to emergency management. Patients with long-term illnesses, patients who have lost their independence, people at risk for accidents, people who need precise daily healthcare follow-up, disabled and elderly people who live alone, older married people with one or both members suffering from a chronic and progressive brain condition (such as Parkinson's or Alzheimer's), individuals who have been subjected to lack of autonomy, pregnant women worried about the prospective danger of miscarriage or early delivery, etc. The HSH system uses cutting-edge technology like electronic systems, intelligent sensors and databases, web-based social networking, and other community projects to protect underprivileged people. The residence could be a private residence owned by a citizen, or it could be a house or a unit inside a hospital [10].

Figure 3.2: In-depth look at the locations and stakeholders engaged in healthcare for the smart home.

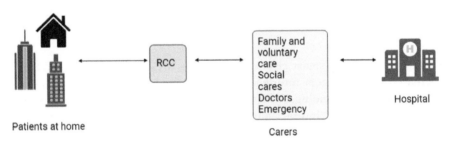

The HSH's integrated approach, which is common in telemedicine technologies, is unique [11]. Unlike the piecemeal approach, the larger HSH notion presents a comprehensive approach to distant care based on the following assumption: telecare technology components operate independently.

Figure 3.3: Conceptual smart-city AI framework.

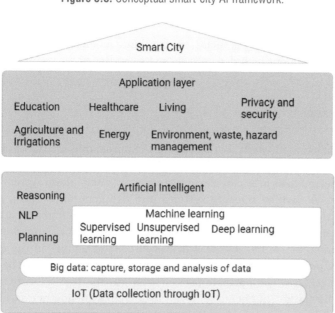

Even for a modest installation, the house will need lots of equipment, including smart sensors with firmware intelligence (firmware is software loaded into a computer's programmed read-only memory (ROM), making it permanent.) Fall detectors are an example of a smart sensors [12]. Certain gadgets, such as alarm clocks or glucose monitors for monitoring the effectiveness of insulin therapy, can be of the therapeutic variety.

Modern "smart houses" have motion-activated lighting, keyless entry systems (usually for the front panel), keyless window and door openers, automatic blinds and drapes, and more [13]. Affluent homeowners who can afford them may consider these gadgets essential. These tools are required for the aged, disabled, and homebound [14]. Given the prevalence of neurological and cognitive deficits in the elderly, installing automated smoke, hazardous gas, microwave, and bath temperature detectors would be advantageous [15]. Renters will feel safer with a home alarm system, which won't raise rent significantly. The MIDAS project [15,16] contains a flood-detector gadget that shuts off bathroom sink water whenever abnormal wetness is detected [17]. These local control solutions should speed up issue resolution to help remote-control the RCC, not automate it. Figure 3.3 shows the smart city AI framework and studies popular smart city applications, as indicated.

3.3 HSH Applications

Telesurveillance for safety is social, while remote medical monitoring is medical. HSH development includes both these two activities.

3.3.1 Social alarm systems

Any basic home equipment that allows for the automated raising of alerts and transmission of those alarms by phone call to an RCC [17] that can offer assistance is referred to as a "social alarm system" (or "safety alarm system" [9,18]). Williams et al. divided telecare systems into two generations [12], although Doughty et al. [10] suggest three generations. Both communities use social alert system prototypes. At this time, the elderly and handicapped employed small, portable alarm triggers to warn loved ones of danger in the case of a fall, illness, predator arrival, or other emergencies [19], and could be a watch or button necklace. The portable press-button was the first telecare device to dominate the market, but home sensors and detectors are now more widespread and accessible. Technology has improved social alert services and systems. Depending on the alarm system, a remote operator will always initiate voice communication with the subscriber at home, check and analyze the situation, look up any relevant information about the subscriber in a database (contact information, location, relevant medical aspects, relatives to contact, etc.), and provide the relevant person with the information they need to respond to a medical or social emergency. Ambulances, home nurses, doctors, police, fire departments, and other support services may be needed. RCC operators use electronic devices to make calls, generate voice contact, and access subscriber databases. They must communicate securely 24/7. Homeowners usually only pay line fees for electronic devices and service membership fees. Due to their easy setup, operation, and maintenance, social alarm systems are more popular than ever. As the market has evolved, and the number of Western private telecare service providers (TSP) have increased, they have become cheaper. Despite this market growth only affecting basic alarm management, this solution uses simple detectors/triggers and a phone communication network. They weren't made for complicated programs like population health monitoring or virtual patient therapy. Automated systems like HSH are expected to replace current safety-alarm systems [18].

3.3.2 Remote patient monitoring

Doughty et al. [10] characterize the second generation of geriatric telecare as remote physiological data monitoring [11]. Sensors and software in this generation track patient data, automatically recognize life-threatening situations, and inform the RCC. Medical research, emergency response, elder care, and chronic illness monitoring have all used these systems [20].

Space medicine is a leader in remote monitoring and intelligent equipment [22]. HSH's long-term success depended on it. Since 1961, NASA has developed guidelines for the space medical community to identify risks, minimize harm, and improve mission success. Mercury (1961–1963) and Gemini (1965–1966) Apollo missions used biosensors to gather and analyze biological signals. Data from 40 missions has been collected. Project Apollo (1968–1972) evaluated crew biosignals, spacecraft interiors, and exteriors. The In-flight Medical Support System (IMSS) is far more advanced than home care. In the 1973 Skylab module, this gadget could alert and diagnose. Medical data monitoring required a novel communication mechanism during Apollo-Soyuz Test Project. NASA Shuttle astronauts can send mission control real-time biological data via the Tracking and Data Relay Satellite System (TDRSS). The Operational Bio-instrumentation System can remotely monitor astronauts' vital signs during EVAs. Future human spaceflights that may require advanced participative and automated medical skills are tested on the ISS. NASA has used ground-based civilian telemedicine since the early 1970s, and remote monitoring of personal and environmental data has proven useful in space medicine [23, 24]. Earthbound patients and space travelers share data collection, transmission, and analysis capabilities. Thus, they can be adapted to Earth life. Smart houses are adding remote monitoring systems that work like social alarms.

3.4 Driving Forces for Health Smart Homes

Due to the surgical economy, individual desires, and joint ventures in cutting-edge technology with a group of investigators from bio- and microsensors and technological means of communication and information processing, houses with smart technology and remote medical assistance have been successfully united.

3.4.1 The medico economic environment

According to Tapscott [25] (used by Tang), the home has taken on a larger role in the healthcare extended enterprise and has developed into a special location for medical treatment and personal care. Public health costs are driven by growing life expectancy. After a second catastrophic fall, more elderly people are moving to a nursing facility. A UK Audit Commission research indicated that over half of hospital beds were taken up by 75-year-olds who could be treated at home.

Remote patient interaction may one day replace in-home medical visits, according to a community nurse study [26]. After this discovery, [27] evaluated the cost of establishing a 20-seat telecare service center to serve a significant number of patients by telephone. Their analysis included heating, hiring more technical staff, and building a new contact center. The attractiveness of maintaining people who have lost autonomy in their homes is demonstrated by numerous case studies in the expert literature in addition to theoretical economic research and other reports.

Modern communication technology can reduce time and effort in patient-nursing-physician-family-volunteer caregiver interactions. Less stress, fewer emergency department visits, shorter hospital stays, quicker discharges, postponed nursing home admissions, and less in-home nursing care are common challenges [10].

3.4.2 Priorities of the individual patient

Telehealthcare's biggest proponent is patients' desire to receive regular medical exams at home [28]. HSH are part of a healthcare movement that stresses patient preferences and requirements alongside medical therapy. Patients are increasingly focusing on therapy and outcomes that are relevant to them [29]. HSHs provide health tracking options, password-protected electronic patient records, and carefully selected medical websites with important medical data. Interactive websites, movies, computer programs, and printed materials are currently available. Tongue excitation and epidermal pressure can help sensory-impaired people communicate. These methods would encounter substantial opposition to adoption even though their sole purpose was to keep people at home, saving money and time. HSHs may help patients' loved ones by helping with repairs or strengthening family bonds. In contrast, patients who receive little medical attention and are given a few smart sensors and software powered by artificial intelligence can feel abandoned and unwanted [30]. Homecare-friendly technology suggests a shift from institutional care for the elderly and disabled to community-based care, which may be cheaper.

3.5 Advanced Technologies

Developments in sensor and information and communication technologies are driving forces behind HSH's sustained expansion. The present technology enables the instrumentation of houses with sensors for the collection of different biosignals and environmental parameters, and network services to the relevant systems for automated alarm triggering. It was inevitable that as communication technologies improved and were more widely used, and as sensors, monitoring, and alarm systems progressed, an integrated system would be deployed in response to the rising demand for citizen telecare. Due to their robust multi-media communication capabilities, internet/intranet networks have become an essential link in the chain of communication for telecare applications. There have been more and more reports of medical systems powered by the Internet [30-32]. While a workable business model for home healthcare systems has not yet been established, the proliferation of start-up companies capitalizing on scientific and technological developments means that these innovations are becoming available to the public at a rapid pace.

3.6 Demonstration Space, Pilot Projects, and Full-Scale Housing

Smart home prototypes and intelligent telemonitoring systems for healthcare and rehabilitation institutions have been developed to meet medico-economic, social, and technical needs. Most have built test sites or exhibition homes, flats, or rooms. A brief, uncritical appraisal of several scientific studies, initiatives, and pilot houses. Due to its breadth and variety, exhaustiveness in this sector is difficult, and commercial exhibition rooms are not covered here.

A UK-based extensive system architecture is CarerNet, and integration of the telecare system is investigated. To formally represent ideas, the writers modified CORE. The result was the creation of a user-centered system model for monitoring tools and intelligent software. The CarerNet concept includes client, home setting, public network, formal caregiver, informal caregiver, emergency services, and service providers. The system model incorporates component needs. The system's intelligent components, distributed among local and remote nodes according to needs, provide robust information-processing capabilities. MIDAS is a second-generation telecare system prototype for the elderly and disabled [15, 16]. The authors say this prototype implements the CarerNet idea. Its ability to detect deviations from a baseline for normal activities and "derive an appraisal of the client's health or status at all times [20]" is significant. Smart, non-intrusive sensors are connected to a local intelligence unit that analyzes data in real time and provides useful technologies like spoken alerts and notifications. A rule-based strategy identifies crises by looking at a series of events and how long each one took.

Based on a basic Greek architecture, the Citizen Health System (CHS) [34] aims to boost IT use in home healthcare across Europe and the world. Its three main goals are IT, healthcare quality, and business. The hospital and the home care services team, which provides home healthcare, make up the system. The first coordinates home healthcare, and is for medical centers and outpatient clinics. The second alternative uses networked wireless in-house sensors. Object-oriented or component-based programming underpins the framework. The program is SCP-ECG and DICOM-compatible and uses the newest Internet technology. Mobile phones and PDAs with WAP capability are among the many cutting-edge technologies used to transport patient data from the remote center to the electronic patient record. Doctors can recommend treatments, distribute health education materials, and perform thorough exams with up-to-date information. Home Asthma Telemonitoring (HAT) [35] provides personalized asthma self-care support. US asthmatics use this Internet-based telemonitoring system which helps healthcare providers quickly identify risky situations. The HAT system transmits all asthma patient data from a doctor's office to their house after further testing. Home asthma telemonitoring lung function tests were comparable to those performed by specialists, and patients were more likely to follow their asthma action plans than those receiving traditional treatment. The system follows patient-centered healthcare model values [35]. The latest research on organizational, behavioral, cognitive, and educational components of asthma self-management guided a multidisciplinary approach to chronic disease management.

3.7 Design of a Smart Home Care System for Rural Elderly Based on AI

12.57% of Chinese people were 65 or older by 2019, and 17.8% needed financial assistance. Rural China's 39.45% of population is always older than urban China. Recently, many rural residents, especially young ones starting families, have moved to cities which has left many rural retirees alone. These retirees can't work. Their remoteness from hospitals, banks, and groceries and lack of reliable transportation causes everyday discomfort and unknown health and safety risks. In the age of AI and 5G, a rural home care system based on AI has successfully addressed many societal concerns affecting the rural old [36]. Smart wearables, AI-enabled Internet connectivity, and intelligent monitoring improve home care. Government agencies and pension service organizations can use user data to increase productivity, help the elderly interact with their children, retain and attract additional consumers, and expand their services. AI enables "intelligent care and quality care" at home [37]. This system relies on an AI that can learn and make decisions [38]. The rural senior home care intelligent system uses sensors to measure environmental conditions and dangerous chemicals. Elderly wearables can track pulse rate, cardiac output, body temperature, sedentary alerts, sleep efficiency, and activity. The smart camera may record the elderly person's movements and behavior in real time to detect early signs of disease, such as falls or mental status changes, and criminal activity. Android workstations can enable intelligent voice chat, phone calls, online product purchases, online medical consultations, and voice alarm services using the man–machine communication module. Big data and cloud computing technologies store, analyze, and process all forms of data, enabling real-time information delivery and early warnings [39]. The rural elderly endowment system uses home-based artificial intelligence technologies for picture and voice recognition. You can use IoT, big data, and cloud computing to monitor and control your home's circumstances, enabling crucial alarms, data analysis, information pushes, early warnings, remote views, and more. Installing Internet of Things (IoT) sensors collects data on indoor temperature, humidity, light levels, and harmful chemicals including carbon monoxide and combustible gases [40]. After installing an intelligent camera, a Raspberry PI application development board with artificial intelligence technology processes, compares, and evaluates target behavior. Smart bracelets collect elders' daily health test results. Elderly folks can use an AI-powered human–computer interaction terminal to join the smart network platform and access a wide range of services, including voice chat, online shopping, and online medical consultations. Environment, health, photo-taking and sharing data are sent wirelessly to the cloud, where they can be preserved, analyzed, and shared. Gateways and interfaces store, evaluate, and push data before sending an early warning and alert if an alarm value exceeds a threshold [40].

3.8 Conclusion

Remote patient care has always intrigued researchers. Monitoring patients at home involves many study disciplines and concerns beyond improving the patient's daily life. Emerging ICTs are used to investigate the best network design and endpoint devices for connecting patients. Medical assistance in identifying whether any physiological and environmental parameters are necessary to assess patients' health, identify suspicious situations, and provide the best medical care; designing an automated, remote-controlled device that could be installed in people's homes could make life more comfortable and safer, inferring health information from sensor data using data fusion, signal processing, and pattern recognition. The sociology and ethics of patient monitoring examine techniques to track patients without invading their privacy. Despite variances, all research outcomes agree to focus telemedicine development on patient needs. Politicians, corporations, and investors desire cheap systems. Remote patient care efforts are scattered and divide projects into those that attempt to improve patients' daily life through automated devices, specific equipment, and basic notifications, and those that assess patients' health and use a global database to provide medical attention to patients outside of a medical setting. Both studies examine separate but related facets of the same issue. Health smart homes will include these perspectives into remote medical treatment. COVID-19 has promoted technology innovation and widespread adoption. Consulting company papers, recent studies, and expert interviews have been reviewed. COVID-19 is affecting the Internet of Things in medicine, transportation, manufacturing, agriculture, housing, and governance.

References

[1]   Stojkoska, B.L.R. and Trivodaliev, K.V. (2017) A review of Internet of Things for smart home: Challenges and solutions. *Journal of Cleaner Production*, 140, 1454–1464.

[2]   Firth, S., Fouchal, F., Kane, T., Dimitriou, V., and Hassan, T. (2013) Decision support systems for domestic retrofit provision using smart home data streams. Available at: https://repository. lboro.ac.uk/articles/conference_contribution/Decision_support_systems _for_domestic_retrofit_provision_using_smart_home_data_streams/ 9425858

[3]   Russell, S.J. (2010) *Artificial Intelligence a Modern Approach*. Pearson Education, Inc.

[4]   Rho, S., Min, G., and Chen, W. (2012) Advanced issues in artificial intelligence and pattern recognition for intelligent surveillance system in smart home environment. *Engineering Applications of Artificial Intelligence*, 25(7), 1299–1300.

[5]   Dermody, G. and Fritz, R. (2019) A conceptual framework for clinicians working with artificial intelligence and health-assistive Smart Homes. *Nursing Inquiry*, 26(1), e12267.

[6]   Kumar, S. and Qadeer, M.A. (2012) Application of AI in home automation. *International Journal of Engineering and Technology*, 4(6), 803.

[7]   Ellis, S. (1999) The patient-centred care model: holistic/multiprofessional/reflective. *British Journal of Nursing*, 8(5), 296–301.

[8]   Vlaskamp, F.J. (1998) SAFE 21-New social alarm services via a proven infrastructure. In *Proc. 3rd TIDE Congres: Technology for Inclusive Design and Equality Improving the Quality of Life for the European Citizen*, 23–25.

[9]   Sixsmith, A.J. (2000) An evaluation of an intelligent home monitoring system. *Journal of Telemedicine and Telecare*, 6(2), 63–72.

[10]   Doughty, K., Cameron, K., and Garner, P. (1996) Three generations of telecare of the elderly. *Journal of Telemedicine and Telecare*, 2(2), 71–80.

[11]   Noury, N., Hervé, T., Rialle, V., Virone, G., Mercier, E., Morey, G., Moro, A., and Porcheron, T. (2000) Monitoring behavior in home using a smart fall sensor and position sensors. In *1st Annual International IEEE-EMBS Special Topic Conference on Microtechnologies in Medicine and Biology. Proceedings (Cat. No. 00EX451)*, 607–610. IEEE.

[12]   Williams, G., Doughty, K., Cameron, K., and Bradley, D.A. (1998) A smart fall and activity monitor for telecare applications. In *Proceedings of the 20th Annual International Conference of the IEEE Engineering in Medicine and Biology Society. Vol. 20 Biomedical Engineering Towards the Year 2000 and Beyond (Cat. No. 98CH36286)*, 3, 1151–1154. IEEE.

[13]   Noury, N., Rialle, V. and Virone, G. (2001) The Telemedecine Home Care Station: a model and some technical hints. *Proc. Healtcomm2001, L'Aquila-Italie*, 37–40.

[14]   Elger, G. and Furugren, B. (1998) SmartBo-an ICT and computer-based demonstration home for disabled people. In *Proceedings of the 3rd TIDE Congress: Technology for Inclusive Design and Equality Improving the Quality of Life for the European Citizen. Helsinki, Finland June (Vol. 1)*.

[15]   Bonner, S.G. (1998) Assisted interactive dwelling house. In *Proc. 3rd TIDE Congress: Technology for Inclusive Design and Equality Improving the Quality of Life for the European Citizen*, 23, 25.

[16]   Rialle, V., Duchene, F., Noury, N., Bajolle, L., and Demongeot, J. (2002) Health "smart" home: information technology for patients at home. *Telemedicine Journal and E-Health*, 8(4), 395–409.

[17]   Celler, B.G., Hesketh, T., Earnshaw, W., and Ilsar, E. (1994) An instrumentation system for the remote monitoring of changes in functional health status of the elderly at home. In *Proceedings of 16th annual international conference of the IEEE engineering in medicine and biology society*, 2, 908–909. IEEE.

[18]   Cooper, M. and Keating, D. (1996) Implications of the emerging home systems technologies for rehabilitation. *Medical Engineering & Physics*, 18(3), 176–180.

[19]   Williams, G., Doughty, K., and Bradley, D.A. (2000) Safety and risk issues in using telecare. *Journal of Telemedicine and Telecare*, 6(5), 249–262.

[20]   Doughty, K., Isak, R., King, P.J., Smith, P., and Williams, G. (1999) MIDAS-Miniature Intelligent Domiciliary Alarm System–a practical application

of telecare. In *Proceedings of the First Joint BMES/EMBS Conference. 1999 IEEE Engineering in Medicine and Biology 21st Annual Conference and the 1999 Annual Fall Meeting of the Biomedical Engineering Society (Cat. N) (Vol. 2, 691-vol).* IEEE.

[21]   Rialle, V., Noury, N., and Hervé, T. (2001) An experimental health smart home and its distributed internet-based information and communication system: first steps of a research project. *Studies in Health Technology and Informatics*, 2, 1479–1483.

[22]   Nicogossian, A.E., Pober, D.F., and Roy, S.A. (2001) Evolution of telemedicine in the space program and earth applications. *Telemedicine Journal and E-health*, 7(1), 1–15.

[23]   Marsh, A. (1998) The Creation of a global telemedical information society. *International Journal of Medical Informatics*, 49(2), 173–193.

[24]   Chen, T.S., Chao, C.M., and Gough, T.G. (1996) Extending an integrated hospital information system beyond the hospital. In *Medical Informatics Europe'96* 680–684. IOS Press.

[25]   Tapscott, D. (1996) *The Digital Economy: Promise and Peril in the Age of Networked Intelligence* New York, NY: McGraw-Hill.

[26]   Mahmud, K. and Lenz, J. (1995) The personal telemedicine system. A new tool for the delivery of health care. *Journal of Telemedicine and Telecare*, 1(3), 173–177.

[27]   Balas, E.A. (1999) Distance technologies for patient monitoring. *BMJ*, 319(7220), 1309.

[28]   Johnson, P. and Andrews, D.C. (1996) Remote continuous physiological monitoring in the home. *Journal of Telemedicine and Telecare*, 2(2), 107–113.

[29]   Ogawa, R. and Togawa, T. (2000) Attempts at monitoring health status in the home. In *1st Annual International IEEE-EMBS Special Topic Conference on Microtechnologies in Medicine and Biology. Proceedings (Cat. No. 00EX451)* 552–556. IEEE.

[30]   Lindberg, C.C. (1997) Implementation of in-home telemedicine in rural Kansas: answering an elderly patient's needs. *Journal of the American Medical Informatics Association*, 4(1), 14–17.

[31]   Rialle, V., Duchene, F., Noury, N., Bajolle, L., and Demongeot, J. (2002) Health "smart" home: information technology for patients at home. *Telemedicine Journal and E-Health*, 8(4), 395–409.

[32]   Noury, N. and Pilichowski, P. (1992) A telematic system tool for home health care. In *Int. Conf. IEEE-EMBS, Paris* 3(7), 1175–1177.

[33]   Nicogossian, A.E., Pober, D.F., and Roy, S.A. (2001) Evolution of telemedicine in the space program and earth applications. *Telemedicine Journal and E-health*, 7(1), 1–15.

[34] Rialle, V., Duchene, F., Noury, N., Bajolle, L., and Demongeot, J. (2002) Health "smart" home: information technology for patients at home. *Telemedicine Journal and E-Health*, 8(4), 395–409.

[35] Peeters, P.H.F. (2000) Design criteria for an automatic safety-alarm system for elderly. *Technology and Health Care*, 8(2), 81–91.

[36] Wang, L.L., Jia, L.Q., Chu, F.Q., and Li, M.X. (2021) Design of home care system for rural elderly based on artificial intelligence. *Journal of Physics: Conference Series*, 1757(1), 012057.

[37] Guoping, D. and Xinshi, C. (2019) Research on the Application Trend of Artificial intelligence in home care under the background of "Internet +" *J. Enterprise Technology and Development* 03 88–89.

[38] Dangchen, S. and Xingchen, L. (2020) Analysis on the applicability of artificial intelligence for home care *J. Journal of Xi'an University of Finance and Economics* 33, 27–36.

[39] Xiaorui, Z., Wenqiong, D., Shijie, P., Weigang, W., and Zhi, L. (2017) Analysis and research on community elderly intelligent medical service system – based on "Internet +" and big data analysis technology *J. Fujian Computer* 33, 144–145.

[40] Hui, X. (2018) Research on multi-sensor Protocol Fusion Technology *D. Shenyang University of Technology*.

Obstacles and Complexity in AI Embedded Remote Health Monitoring

Abstract

AI may solve several healthcare system issues worldwide. Implementation and innovation studies show that healthcare executives oppose new technology, slowing and varying their adoption. Despite substantial research on other stakeholders, few studies have examined how healthcare leaders feel about AI deployment. Healthcare executives' perspectives are essential since they help implement new technology in healthcare delivery. Precision medicine could improve symptom-driven treatment by allowing early interventions, improving diagnostics, and producing better, cheaper treatments. Both patient data and general factors that can track and distinguish healthy and unwell people must be examined to determine the best approach to personalized and population therapy. This may help us understand health-related biochemical markers. Despite technological advances, applying healthcare data to diagnosis and treatment planning is difficult due to each patient's condition intricacy. Precision medicine improves patient outcomes and provides real-time decision assistance by integrating previously segregated data sources and discovering disease progression patterns unique to each patient. We require analytical tools, technologies, databases, and techniques to connect and interoperate clinical, laboratory, and public health information systems to better balance ethical and social issues related to healthcare data privacy and protection. Digital healthcare may enhance access and flexibility. It covers biological research, health, treatment, and issues. Low-income countries offer diagnostic and medical services. Digital health technology struggles with reliability, safety, testing, and ethics.

4.1 Introduction

In the past decade, artificial intelligence (AI) has advanced significantly. Evaluation of thousands of medical records helps speed up and improve treatment. AI simulates human intellect and processing speed by training a computer to operate like a brain. This technology quickly acquires knowledge, makes predictions, analyzes, draws conclusions, and corrects course. These systems are used for imaging, problem solving, voice recognition, and skill acquisition. To create more accurate predictions and help solve complex problems, artificial intelligence systems can be educated on a dataset [1-3]. Artificial intelligence can save healthcare providers time by digitizing patient records and organizing them into a database for diagnosis, treatment, and Medicare billing. Medical professionals must engage with programmers and engineers to create a system for data collection and monotonous tasks that meets all requirements. There's a tendency toward customizing general-purpose programs for specific businesses. Doctors and surgeons have helped AI grow more creative. Smart computers quickly understand medical data, financial transaction jargon, text, and photographs. Without human input, these machines can understand and make the right judgment [4, 5]. The patient receives clear instructions from the guideline. Using this technology, we may categorize people, gather data, and use this knowledge to foresee and prevent probable difficulties during joint replacement surgery, shorten patients' hospital stays, and improve their chances of a full recovery [6, 7]. AI is the best way to prolong life. AI-assisted robot surgery is tough. This technology creates information by regularly interacting with the patient and using interactive tools. Rural areas lack medical experts, but new technologies may assist. The initiative helps prospective doctors fulfill unanticipated needs in underdeveloped areas [8, 9]. This technology helps doctors work quicker and better, and patients get better treatment at a reduced cost, thus, it helps doctors diagnose [10, 11]. CT, MRI, X-ray, and 3D scanners use AI, and the patient has more enticing options. AI advises a balanced diet, it's a terrific way to keep track of patient appointments, and it has been successful in medical settings for remote doctor–patient communication [12-14].

4.2 Healthcare With AI

AI is used in healthcare to improve early disease detection, comprehend disease evolution, optimize drug and therapy dosages, and develop new medicines.

Artificial intelligence is best at analyzing massive datasets quickly. These fields of medicine are particularly promising for pattern recognition because artificial intelligence (AI) systems can study photographs to discover anomalies or subtleties that the human eye might overlook (such as gender from the retina) as good as or better than experienced physicians [15, 16]. AI may never replace doctors, but smart technology is improving clinical decision-making. AI-enabled robots can quickly synthesize medical information from countless sources, but people are restricted by their willingness to learn new information, abilities, and experience [17]. Using large datasets like electronic health records to computationally analyze human actions and routines can advance artificial intelligence (AI) [18-20]. AI may help biological research and clinical trials, will aid automation and hard manual labor, but despite recent advances, AI cannot yet replace humans in health and biology research. Telemedicine, blockchain, "big data," artificial intelligence, and other cutting-edge technologies are being employed to address healthcare and medical education concerns.

4.3 The Positive Effects of AI in Healthcare

- Screening for irregularities and recommend medical intervention
- Predicting future illnesses
- Precise and successful diagnosis
- Useful for nuanced and new treatments
- Balancing the patient's blood/glucose levels
- Adequate patient care
- Providing comfort for physicians and patients
- Proper preparation for students of medicine
- Improving hospital protection
- During surgery, gathering data to help develop the potential procedure
- Good results for patients
- Improving the experience of doctors/surgeons
- Diagnostic results
- Enhanced pathological outcome
- Reduced diagnostic costs
- Holding a clinical record
- Providing the patient with outstanding care.

Healthcare AI can measure carina angle, aortic valve, and pulmonary artery diameter, and it evaluates orthopaedic fracture and trauma [21, 22]. It may also aid global healthcare [23-25]. A computer system with "artificial intelligence" (AI) can do tasks and reason like humans [26]. Thus, AI encompasses medical diagnosis and therapy, patient education and compliance, and back-office duties [27, 28]. Optimization may help implement AI technologies. Clinical healthcare AI may require complex sociotechnical modifications [29]. Artificial intelligence can help diagnose and cure many medical problems, aid in decision-making, reduce patient care errors, boost efficiency and accessibility, improve patient experiences and outcomes, and lower per capita healthcare costs [30-32]. AI could enhance healthcare, but it hasn't [24, 27, 33].

Cancer, neurology, and cardiology dominate AI research [24, 25, 32, 34, 35]. We disagree on how long it takes to develop trustworthy methods and apply AI systems to healthcare. Recently there have been studies on regulatory monitoring, privacy, and legal ramifications [36, 37], ethics [38, 39], healthcare delivery, patient health, and financial resources [40-42] in artificial systems. Real-world clinical investigations are needed because clinical AI application is new [43] and few AI technology implementation frameworks or models exist [44].

This indicates the vast gap in understanding between healthcare executives, educating both healthcare employees and patients on the benefits of AI in practice, and the variety of views on its acceptance [35]. Healthcare executives distrust AI [34, 45-47] and because of its compatibility and impact on work processes, healthcare administrators rarely install and use new technology [48]. Healthcare AI studies are sparse.

Healthcare AI requires AI-savvy leaders. Leaders must promote artificial intelligence (AI) technology, figure out how to incorporate it without disturbing operations, and examine how AI might increase productivity, patient safety, and healthcare access [49, 50]. Leadership drives culture, performance, and innovation [51–53]. Many implementation study paradigms, models, and concepts emphasize leadership in propagating new healthcare practices [54]. The Consolidated Framework for Implementation Research (CFIR), which includes dynamic implementation frameworks and adapted implementation for long-term illnesses, states that leadership is essential for successful implementation. Healthcare researchers employ these models, which aren't AI-ready.

4.4 Using AI to Improve Healthcare in Areas Such as Illness Prevention, Detection, and Treatment

If they accepted technological progress, healthcare professionals could gain from artificial intelligence. They could focus on eliminating tedious tasks with AI to improve doctor–patient interactions and to increase empathy and IQ. The authors prioritized data flow because it allows computers to independently design a sophisticated function with improved prediction given a lot of input. They developed an image processing technique for diabetic retinopathy and

Figure 4.1: An illustration of the main uses of digital technology in healthcare.

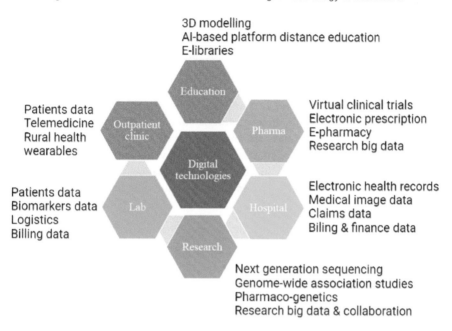

3D modelling
AI-based platform distance education
E-libraries

Patients data
Telemedicine
Rural health
wearables

Virtual clinical trials
Electronic prescription
E-pharmacy
Research big data

Patients data
Biomarkers data
Logistics
Billing data

Electronic health records
Medical image data
Claims data
Biling & finance data

Next generation sequencing
Genome-wide association studies
Pharmaco-genetics
Research big data & collaboration

a massive neural network for skin cancer [55, 56]. They have developed a smartphone-accessible AI tool to assess direct oral drug compliance. The project could improve medical access in underprivileged areas by allowing primary care providers to use AI to prescribe pharmaceuticals even when specialists aren't available. The possibility of patient data being misused and the resulting loss of privacy, however, is not being appropriately addressed, although there is the possibility of HIPAA (Health Insurance Portability and Accountability Act) compliance being required[74, 75]. Recent advancements and developments in the application of digital health technology, AI, "big data," telemedicine, and blockchain technology are employed to address major healthcare and medical education concerns (Figure 4.1).

4.5 Background Theory for AI and ML, with Instances in the Healthcare Industry

Artificial intelligence allows computers to understand patterns of interaction among variables, learn from their failures, devise plans, and predict better future actions. AI continues to interest organizations of all sizes and sectors nearly 50 years after its invention [59, 60]. Machine translation, natural language processing, data mining, risk modeling, image recognition, machine vision, knowledge bases, and agent-based systems can be used in computational command line, desktop, web, robotic, and mobile applications. Reactive machines, limited memory, theory of mind, and self-aware machines are the four main artificial intelligence categories that can be broken down into how decisions are made (based on human-level consciousness). Over the last few years, AI has been used to analyze clinical data (electronic health records, images, etc.) to improve diagnosis and treatment in a variety of fields [61], including radiology [62], early diagnosis, improved visualization of diseases, and emergency scenario prediction [63, 64], oncology (diagnostic of breast [65], skin [66], and lung [67] cancer), and others. Precision medicine analyzes patient data using ML algorithms [68-70].

4.6 Improved Patient Care via Machine Learning and Increased Provider–Patient Communication

This method investigates the basic healthcare system changes needed to realize the medical potential of machine learning (ML). It focuses ML in medicine, which may combine historical data with expert opinion to customize a patient's diagnosis and treatment. The authors discussed how difficult it is to combine the results of modern ML models with those of traditional statistical models, how much data is needed, the need to train machine learning classifiers for identifying broad and subtle relationships, and the urgency of educating medical professionals on this topic, using the application of artificial intelligence to thoroughly comprehend data. EHRs may squander doctors' time on administrative and billing tasks (such as check boxes), and doctors are concerned that EHR ML would make clinicians less aware of errors and algorithmic bias. EHRs should train an ML classification model for pattern recognition to predict high-risk patients' outcomes, diagnose them, and provide a fast search engine for crucial patient data [71].

4.7 Oncology, Big Data, and AI

Cognition-enabling computing (read, recall, propose, and remind) in cancer diagnosis and treatment appears promising (read, remember, recommend and remind). The authors of research suggest that cognitive computer systems might help medical professionals by simplifying their access to relevant data and established practices. Such platforms can spread cancer information into clinical practice and around the world by improving clinical research and trials with less bureaucracy and cost. AI could create global cancer networks, develop therapeutic approaches for rare and aggressive cancers, observe therapeutic outcomes by different parameters, analyze associations between cancer and other disease-specific attributes, advance new cancer etiologies, and more. Large AI data repositories may also aid cancer research and therapy [72, 73].

4.8 Intelligent Digital Pathology with Deep Learning

Deep learning can improve medical diagnosis by analyzing whole-slide pathology photos, which are hard to see. Deep learning was tested on 129 whole-slide photos of metastatic tissue slices from breast cancer patients' lymph nodes. The researchers beat 11 pathologists by fine-tuning deep learning systems. The researchers were optimistic about digital pathology applications of deep learning that highlighted the need for new intelligence-based gear to address diagnostic sensitivity and specificity [74].

4.9 Smart Health Data Analytics for Better Healthcare System Administration, Patient Outcomes, Scientific Discovery, and New Treatment Methods

Since many healthcare systems create large amounts of heterogeneous data, it must be leveraged to improve resource efficiency, patient satisfaction, treatment quality, and health outcomes. Advanced health data analytics based on AI and ML can provide non-linear connections and causal correlations between intrinsic data items to enable customized and prescriptive medication by understanding the arrangement of active partners in healthcare processes. Healthcare information analytics researchers employ data- and knowledge-driven methods to make decisions, describe data, optimize processes, and do comparative, prescriptive, and semantic analyses. They recommended using AI and ML to pre-process health data, choose an algorithm based on the desired outcome, build analytical models, and extrapolate data importance [75].

4.10 Artificial Intelligence (AI) Applications in Orthopedics

Orthopedics can benefit from AI: human-like AI knowledge, reason, self-correction, robotics, and other technology ensure orthopedic surgical precision [76]. These gadgets detect errors and report heat, light, motion, temperature, sound, and pressure. AI excels at real-time monitoring, management, and communication. Software and computers may now replicate human intellect [77], helping clinicians provide efficient and individualized care [78]. AI solves the personalization problem by delivering specific patient data, boosting health and welfare.

Orthopedic surgeons use AI to identify and treat bone cancer, traumatic bone fractures, arthroplasty, and bone marrow diseases as it can assess patient health, function, and quality of life [4]. Table 4.1 lists orthopedic AI applications. AI can read X-ray, CT, and MRI data (MRI), and it diagnoses and treats musculoskeletal trauma. This method improves orthopedic surgeon precision from 83% to 97% [79]. AI trains doctors, saving lives, and it could improve orthopedic and trauma surgery. AI may design and deploy effector weapons that change behavior [80].

Table 4.1: Uses of AI in the field of orthopedics.

S. NO	Application	Description
1	Skeletal radiograph analysis	AI can highlight anatomical characteristics in a skeletal radiograph
2	Surgical training	Compares fracture healing following therapy Surgical training is possible without live patients
3	Coordination and treatment	Diagnostic, therapeutic, and treatment data are stored Patient–surgeon communication
4	Increased proficiency in orthopedic robot-assisted surgery	Surgeons' performance may be assessed and improved during surgery Improves robot-assisted surgery by optimizing surgical equipment
5	Lessen hospital and follow-up visits	It improves the results of surgery and shortens the length of time patients have to stay in the hospital AI delivers patient-related information to decrease number of hospital visits
6	Problem-solving	AI improves medical diagnosis, therapy, and surgery It lets medical devices think, learn, communicate, and solve issues

The superior intelligence provided by machine/information systems that can forecast, perform, analyze, evaluate, and verify a preset setting is artificial intelligence (AI). This technique accurately predicts and controls post-operative infections [81]. To improve surgical efficiency and patient safety, orthopedic surgeons may now make use of cutting-edge technology that allows them to record relevant data during procedures. The surgical process becomes more intelligent and effective as a result [82]. The current user of AI must bear in mind the limits that are intrinsic to the technology (Table 4.2).

Table 4.2: Limitations of artificial intelligence.

S. No	Limitation	Description
1	Only uses the input data to learn	AI uses patient data to learn Accuracy of surgery relies on data
2	Relevant to the algorithm's decision to treat with ML/AI	ML/AI system prediction accuracy, essential for therapy and beyond, is a restriction The outcome of the prediction will show whether AI plays an absolute role or if it needs more data and training over time
3	Does not comprehend feelings	These robots don't grasp emotions, human cognition, or reasons for making decisions This technology can only carry out the tasks for which it has been designed It has no other knowledge like a human being to reflect the accurate result
4	Supervision needed	Doctors and surgeons can't get good outcomes unless they have close monitoring and record everything that happens throughout therapy Only accurate data produces outcomes
5	Does not provide creative thinking	It can't replace a human surgeon's creative thinking in orthopedics since people have the capacity for emotion, reason, and discernment that a machine lacks This technology cannot be used to make new decisions in the absence of information

Artificial intelligence could improve bone loss detection, treatment planning, imaging, and diagnosis in a computerized information system. AI improves patient–physician communication, surgical efficiency, and complication risks during difficult surgeries. Better information and communication channels should boost efficiency and patient satisfaction. It aids medical training and AI will aid orthopedic surgeons and other medical professions [83]. AI uses data to solve challenging issues in a creative way and helps choose surgical implants for personalized surgical case plans. Operations, clinical, and surgical management may benefit patients. This technology may benefit orthopedic, spinal, and trauma surgery. Thus, AI can improve orthopedic surgery and patient care.

4.10.1 Telemedicine

With the use of communication technology, healthcare professionals may examine, diagnose, and treat patients in distant areas thanks to telemedicine [84, 85]. Collecting, storing, and exchanging medical data is one of telemedicine's

Figure 4.2: The application of telemedicine in healthcare.

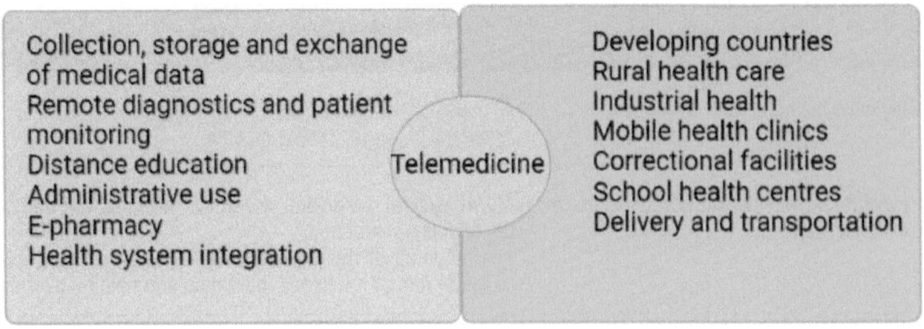

Collection, storage and exchange of medical data Remote diagnostics and patient monitoring Distance education Administrative use E-pharmacy Health system integration	Developing countries Rural health care Industrial health Mobile health clinics Correctional facilities School health centres Delivery and transportation

Telemedicine

benefits [86]. Additionally, telemedicine enables patient mobility tracking, remote patient monitoring, distant learning, better healthcare administration and management, and integration of health data systems [87, 88]. In reality, there are several opportunities for telemedicine use across different industries (Figure 4.2).

4.10.2 Rural healthcare

People in rural areas needing access to medical care in remote areas is one of the main problems in rural healthcare [95, 96]. In this situation, telemedicine may aid in the resolution of these issues by enabling access to high-quality medical treatments wherever the patient may be. Using customized online services or in-the-moment video conferencing are two options. The physicians can communicate with the patient and provide guidance thanks to a mix of medical technology, television, and specialized wearables [88]. The doctor may examine the patient, take some important vital signs, go over their medical history, assess, diagnose, and then decide on a course of therapy and prescribe medication. A rapid line of communication and patient input are made possible by such a method. Furthermore, it may greatly reduce the need for very costly commuting, which is crucial for areas with harsh climate conditions.

4.11 Conclusion

In conclusion, healthcare professionals have highlighted internal and external barriers to AI adoption. Evolution of healthcare professions and methods, as well as external factors and the capacity for strategic change management have been problems. Our findings suggest that healthcare organizations at all levels need to view AI implementation as an ongoing learning process, one that calls for improved systems thinking. The field of precision medicine is making strides, but it still faces a number of obstacles. This includes a need for more efficiency to continue enhancing the connectedness and interoperability of clinical, laboratory, and public health systems, and to address ethical and social challenges linked to the privacy and protection of healthcare and omics data. The intelligent processing of huge amounts of structured clinical data is one of the possible benefits that could result from the creation of new AI and ML-based big data platforms. Such platforms have the ability to transform medicine, and hence improve the overall quality of healthcare and facilitate its transformation.

References

[1] Haleem, A., Vaishya, R., Javaid, M., and Khan, M. I (2019) Artificial Intelligence (AI) applications in orthopaedics: An innovative technology to embrace. *Journal of Clinical Orthopaedics and Trauma.* https://doi.org/10.1016/j.jcot.2019.06.012

[2] Jha, S. and Topol E. J. (2018) Information and artificial intelligence. *J Am Coll Radiol* 15(3), 509–511.

[3] Lupton, M. (2018) Some ethical and legal consequences of the application of artificial intelligence in the field of medicine. *Trends in Medicine* 18(4), 1-7.

[4] Murdoch, T. B. and Detsky, A.S. (2013) The inevitable application of big data to health care. *JAMA* 309(13), 1351–1352.

[5] Misawa, M., Kudo, S. E., Mori, Y., et al. (2018) Artificial intelligence-assisted polyp detection for colonoscopy initial experience. *Gastroenterology* 154, 2027–2029.

[6] Caocci, G., Baccoli, R., Vacca, A., Mastronuzzi, A., Bertaina, A., Piras, E., Littera, R., Locatelli, F., Carcassi, C., and La Nasa, G. (2010) Comparison between an artificial neural network and logistic regression in predicting acute graft-vs-host disease after unrelated donor hematopoietic stem cell transplantation in thalassemia patients. *ExpHematol.* 38, 426–433.

[7] Haleem, A., Javaid, M., and Vaishya, R. (2019) Industry 4.0 and its applications in orthopaedics. *J ClinOrthop Trauma* 10(3), 615-616.

[8] Guo, J. and Li, B. (2018) The application of medical artificial intelligence technology in rural areas of developing countries. *Health Equity* 2(1), 174–181.

[9] Atasoy, H., Greenwood, B. N., and McCullough, J. S. (2018) The digitization of patient care: A review of the effects of electronic health records on health care quality and utilization. *Annu Rev Public Health* 40, 1. doi: 10.1146/annurev-pub health-040218-044206.

[10] Jiang, F., Jiang, Y., Zhi, H., Dong, Y., Li, H., Ma, S., Wang, Y., Dong, Q., Shen, H., and Wang, Y. (2017) Artificial intelligence in healthcare: past, present and future. *Stroke and Vascular Neurology* 2, 230–43.

[11] Haleem, A., Javaid, M., Haleem, A., and Javaid, M. (2019) Industry 5.0 and its expected applications in medical field. *Current Medicine Research and Practice* 9(4), 167-169.

[12] Buch, V. H., Ahmed, I., and Maruthappu, M. (2018) Artificial intelligence in medicine: current trends and future possibilities. *Br J Gen Pract* 68(668), 143-144.

[13] Kulikowski, C. A. (2019) Beginnings of artificial intelligence in medicine (AIM): computational artifice assisting scientific inquiry and clinical art - with reflections on present AIM challenges. *Yearb Med Inform.* doi: 10.1055/s-0039-1677895.

[14] Upadhyay, A. K. and Khandelwal, K. (2019) Artificial intelligence-based training learning from application. *Development and Learning in Organizations: An International Journal* 33(2), 20-23.

[15] Brinker, T. J., Hekler, A., Hauschild, A., Berking, C., Schilling, B., Enk, A. H., et al. (2019) Comparing artificial intelligence algorithms to 157 German dermatologists: the melanoma classification benchmark. *Eur J Cancer* 111, 30–7. https://doi.org/10.1016/j.ejca.2018. 12.016.

[16] Hosny, A., Parmar, C., Quackenbush, J., Schwartz, L. H., and Aerts, H. J. W. L. (2018) Artificial intelligence in radiology. *Nat Rev Cancer* 18(8), 500–10. https://doi.org/10.1038/s41568-018-0016-5.

[17] Sengupta, P. P. and Adjeroh, D. A. (2018) Will artificial intelligence replace the human echocardiographer? *Circulation* 138(16), 1639–42. https://doi.org/10.1161/CIRCULATIONAHA.118.037095.

[18] Vidal-Alaball, J., RoyoFibla, D., Zapata, M. A., Marin-Gomez, F. X., and Solans, F. O. (2019) Artificial intelligence for the detection of diabetic retinopathy in primary care: protocol for algorithm development. *JMIR Res Protoc.* 8(2), e12539. https://doi.org/10.2196/ 12539.

[19] Topol, E. (2019) Deep Medicine: How Artificial Intelligence Can Make Healthcare Human Again, 1st edn. New York, NY: Basic Books.

[20] Wang, Y., Kung, L. A., Byrd, T. A. (2016) Big data analytics: understanding its capabilities and potential benefits for healthcare organizations. *Technol Forecast Soc Change* 126, 3–13. https://doi.org/ 10.1016/j.techfore.2015.12.019

[21] Lee, E. J., Kim, Y. H., Kim, N., et al. Deep into the brain: artificial Intelligence in stroke imaging. *J Stroke* 19, 277–285.

[22] Muhsen, I. N., El Hassan, T., and Hashmi, S. K. (2018) Artificial intelligence approaches in hematopoietic cell transplantation: A review of the current status and future directions. *Turk J Haematol.* 35(3), 152-157.

[23]   Buch, V. H., Ahmed, I., and Maruthappu, M. (2018) Artificial intelligence in medicine: current trends and future possibilities. *British Journal of General Practice* 68(668), 143-144.

[24]   Mehta, N., Pandit, A., and Shukla, S. (2019) Transforming healthcare with big data analytics and artificial intelligence: A systematic mapping study. *Journal of Biomedical Informatic*, 100, 103311.

[25]   Horgan, D., Romao, M., Morré, S.A. and Kalra, D. (2019) Artificial intelligence: power for civilisation–and for better healthcare. *Public Health Genomics* 22(5-6), 145-161.

[26]   European Commission (2018) *A Definition of AI: Main Capabilities and Scientific Disciplines*. Brussels: European Commission.

[27]   Davenport, T. and Kalakota, R. (2019) The potential for artificial intelligence in healthcare. *Future Healthcare Journal* 6(2), p.94.

[28]   Shaw, J., Rudzicz, F., Jamieson, T., and Goldfarb, A. (2019) Artificial intelligence and the implementation challenge. *Journal of Medical Internet Research* 21(7), p.e13659.

[29]   Elish, M.C. (2018) The stakes of uncertainty: developing and integrating machine learning in clinical care. In *Ethnographic Praxis in Industry Conference Proceedings* 2018(1), 364-380.

[30]   Lee, J.C. (2019) The perils of artificial intelligence in healthcare: Disease diagnosis and treatment. *Journal of Computational Biology and Bioinformatics Research* 9(1), 1-6.

[31]   Topol, E.J. (2019) High-performance medicine: the convergence of human and artificial intelligence. *Nature Medicine* 25(1), 44-56.

[32]   Jiang, F., Jiang, Y., Zhi, H., Dong, Y., Li, H., Ma, S., Wang, Y., Dong, Q., Shen, H., and Wang, Y. (2017) Artificial intelligence in healthcare: past, present and future. *Stroke and Vascular Neurology* 2(4).

[33]   Petersson, L., Larsson, I., Nygren, J.M., Nilsen, P., Neher, M., Reed, J.E., Tyskbo, D., and Svedberg, P. (2022) Challenges to implementing artificial intelligence in healthcare: a qualitative interview study with healthcare leaders in Sweden. *BMC Health Services Research* 22(1), 1-16.

[34]   He, J., Baxter, S.L., Xu, J., Xu, J., Zhou, X., and Zhang, K. (2019) The practical implementation of artificial intelligence technologies in medicine. *Nature Medicine* 25(1), 30-36.

[35]   Alhashmi, S.F., Alshurideh, M., Kurdi, B.A., and Salloum, S.A. (2020) A systematic review of the factors affecting the artificial intelligence implementation in the health care sector. In *The International Conference on Artificial Intelligence and Computer Vision*. Cham: Springer, 37-49.

[36]   Asan, O., Bayrak, A.E., and Choudhury, A. (2020) Artificial intelligence and human trust in healthcare: focus on clinicians. *Journal of medical Internet research* 22(6), e15154.

[37]   Gooding, P. and Kariotis, T. (2021) Ethics and law in research on algo-
       rithmic and data-driven technology in mental health care: scoping
       review. *JMIR Mental Health* 8(6), e24668.

[38]   Beil, M., Proft, I., van Heerden, D., Sviri, S., and van Heerden, P.V. (2019)
       Ethical considerations about artificial intelligence for prognostication in
       intensive care. *Intensive Care Medicine Experimental* 7(1), 1-13.

[39]   Murphy, K., Di Ruggiero, E., Upshur, R., Willison, D.J., Malhotra, N., Cai,
       J.C., Malhotra, N., Lui, V., and Gibson, J. (2021) Artificial intelligence
       for good health: a scoping review of the ethics literature. *BMC Medical
       Ethics* 22(1), 1-17.

[40]   Choudhury, A. and Asan, O. (2020) Role of artificial intelligence in
       patient safety outcomes: systematic literature review. *JMIR Medical
       Informatics* 8(7), e18599.

[41]   Fernandes, M., Vieira, S.M., Leite, F., Palos, C., Finkelstein, S., and Sousa,
       J.M. (2020) Clinical decision support systems for triage in the emergency
       department using intelligent systems: a review. *Artificial Intelligence in
       Medicine* 102, 101762.

[42]   Yin, J., Ngiam, K.Y. and Teo, H.H. (2021) Role of artificial intelligence
       applications in real-life clinical practice: systematic review. *Journal of
       Medical Internet Research* 23(4), e25759.

[43]   Wolff, J., Pauling, J., Keck, A., and Baumbach, J. (2020) The economic
       impact of artificial intelligence in health care: systematic review. *Journal
       of Medical Internet Research* 22(2), p.e16866.

[44]   Gama, F., Tyskbo, D., Nygren, J., Barlow, J., Reed, J., and Svedberg, P.
       (2022) Implementation frameworks for artificial intelligence translation
       into health care practice: Scoping review. *Journal of Medical Internet
       Research* 24(1), p.e32215.

[45]   Safi, S., Thiessen, T., and Schmailzl, K.J. (2018) Acceptance and resis-
       tance of new digital technologies in medicine: qualitative study. *JMIR
       Research Protocols* 7(12), e11072.

[46]   Whitelaw, S., Mamas, M.A., Topol, E., and Van Spall, H.G. (2020) Appli-
       cations of digital technology in COVID-19 pandemic planning and
       response. *The Lancet Digital Health* 2(8), e435-e440.

[47]   Alami, H., Lehoux, P., Denis, J.L., Motulsky, A., Petitgand, C., Savoldelli,
       M., Rouquet, R., Gagnon, M.P., Roy, D., and Fortin, J.P. (2020) Orga-
       nizational readiness for artificial intelligence in health care: insights
       for decision-making and practice. *Journal of Health Organization and
       Management* Preprint.

[48]   Reed, J.E., Howe, C., Doyle, C., and Bell, D. (2018) Simple rules for
       evidence translation in complex systems: a qualitative study. *BMC
       Medicine* 16(1), 1-20.

[49]   Chen, M. and Decary, M. (2020) Artificial intelligence in health-care: An essential guide for health leaders. In *Healthcare Management Forum* 33(1), 10-18.

[50]   Loh, E. (2018) Medicine and the rise of the robots: a qualitative review of recent advances of artificial intelligence in health. *BMJ Leader* 2(2).

[51]   Ogbonna, E. and Harris, L.C. (2000) Leadership style, organizational culture and performance: empirical evidence from UK companies. *International Journal of Human Resource Management* 11(4), 766-788.

[52]   Battilana, J., Gilmartin, M., Sengul, M., Pache, A.C., and Alexander, J.A. (2010) Leadership competencies for implementing planned organizational change. *The Leadership Quarterly* 21(3), 422-438.

[53]   Denti, L. and Hemlin, S. (2012) Leadership and innovation in organizations: A systematic review of factors that mediate or moderate the relationship. *International Journal of Innovation Management* 16(03), 1240007.

[54]   Nilsen, P. (2020) Overview of theories, models and frameworks in implementation science. In *Handbook on Implementation Science* Edward Elgar Publishing, 8-31.

[55]   Bailey, F. W. (2012) Key concepts, themes, and evidence for practitioners in educational psychology. *Handbook of Implementation Science for Psychology in Education* Cambridge University Press, 13.

[56]   Flottorp, S. A., Oxman, A. D., Krause, J., Musila, N. R., Wensing, M., Godycki-Cwirko, M., Baker, R., and Eccles, M. P. (2013) A checklist for identifying determinants of practice: a systematic review and synthesis of frameworks and taxonomies of factors that prevent or enable improvements in healthcare professional practice. *Implementation Science* 8(1), 1-11.

[57]   Gulshan, V., Peng, L., Coram, M., Stumpe, M. C., Wu, D., Narayanaswamy, A., Venugopalan, S., Widner, K., Madams, T., Cuadros, J., and Kim, R. (2016) Development and validation of a deep learning algorithm for detection of diabetic retinopathy in retinal fundus photographs. *JAMA* 316(22), 2402-2410.

[58]   Labovitz, D. L., Shafner, L., Reyes Gil, M., Virmani, D., and Hanina, A. (2017) Using artificial intelligence to reduce the risk of nonadherence in patients on anticoagulation therapy. *Stroke* 48(5), 1416-1419.

[59]   Ahmed, Z., Kim, M., and Liang, B. T. (2019) MAV-clic: management, analysis, and visualization of clinical data. *JAMIA Open* 2(1), 23-28.

[60]   Ahmed, Z. and Liang, B. T. (2019) Systematically dealing practical issues associated to healthcare data analytics. In *Future of Information and Communication Conference* Cham: Springer, 599-613)..

[61]   Buch, V. H., Ahmed, I. and Maruthappu, M. (2018) Artificial intelligence in medicine: current trends and future possibilities. *British Journal of General Practice* 68(668), 143-144.

[62]   Bali, J., Garg, R., and Bali, R. T. (2019) Artificial intelligence (AI) in healthcare and biomedical research: Why a strong computational/AI bioethics framework is required?. *Indian Journal of Ophthalmology* 67(1), 3.

[63]   Mintz, Y. and Brodie, R. (2019) Introduction to artificial intelligence in medicine. *Minimally Invasive Therapy & Allied Technologies* 28(2), 73-81.

[64]   Gulshan, V., Peng, L., Coram, M., Stumpe, M.C., Wu, D., Narayanaswamy, A., Venugopalan, S., Widner, K., Madams, T., Cuadros, J., and Kim, R. (2016) Development and validation of a deep learning algorithm for detection of diabetic retinopathy in retinal fundus photographs. *JAMA* 316(22), 2402-2410.

[65]   Bejnordi, B. E., Veta, M., Van Diest, P. J., Van Ginneken, B., Karssemeijer, N., Litjens, G., Van Der Laak, J. A., Hermsen, M., Manson, Q. F., Balkenhol, M., and Geessink, O. (2017) Diagnostic assessment of deep learning algorithms for detection of lymph node metastases in women with breast cancer. *JAMA* 318(22), 2199-2210.

[66]   Prevedello, L. M., Erdal, B. S., Ryu, J. L., Little, K. J., Demirer, M., Qian, S., and White, R. D. (2017) Automated critical test findings identification and online notification system using artificial intelligence in imaging. *Radiology* 285(3), 923-931.

[67]   Somashekhar, S. P., Sepúlveda, M. J., Puglielli, S., Norden, A. D., Shortliffe, E. H., Kumar, C. R., Rauthan, A., Kumar, N. A., Patil, P., Rhee, K., and Ramya, Y. (2018) Watson for Oncology and breast cancer treatment recommendations: Agreement with an expert multidisciplinary tumor board. *Annals of Oncology* 29(2), 418-423.

[68]   Esteva, A., Kuprel, B., Novoa, R. A., Ko, J., Swetter, S. M., Blau, H. M,. and Thrun, S. (2017) Dermatologist-level classification of skin cancer with deep neural networks. *Nature* 542(7639), 115-118.

[69]   Galimova, R. M., Buzaev, I. V., Ramilevich, K. A., Yuldybaev, L. K., and Shaykhulova, A.F. (2019) Artificial intelligence-developments in medicine in the last two years. *Chronic Diseases and Translational Medicine* 5(01), 64-68.

[70]   Mesko, B. (2017) The role of artificial intelligence in precision medicine. *Expert Review of Precision Medicine and Drug Development* 2(5), 239-241.

[71]   Van Hartskamp, M., Consoli, S., Verhaegh, W., Petkovic, M., and Van de Stolpe, A. (2019) Artificial intelligence in clinical health care applications. *Interactive Journal of Medical Research* 8(2), e12100.

[72]   Schork, N. J. (2019) Artificial intelligence and personalized medicine. In *Precision Medicine in Cancer Therapy* Cham: Springer, 265-283.

[73]   Rajkomar, A., Dean, J., and Kohane, I. (2019) Machine learning in medicine. *New England Journal of Medicine* 380(14), 1347-1358.

[74]   Stein, J. D., Rahman, M., Andrews, C., Ehrlich, J. R., Kamat, S., Shah, M., Boese, E. A., Woodward, M. A., Cowall, J., Trager, E. H., and Narayanaswamy, P. (2019) Evaluation of an algorithm for identifying ocular conditions in electronic health record data. *JAMA Ophthalmology* 137(5), 491-497.

[75]   Hinton, G. (2018) Deep learning—a technology with the potential to transform health care. *JAMA* 320(11), 1101-1102.

[76]   Acs, B. and Rimm, D.L. (2018) Not just digital pathology, intelligent digital pathology. *JAMA Oncology* 4(3), 403-404.

[77]   Olczak, J., Fahlberg, N., Maki, A., Razavian, A. S., Jilert, A., Stark, A., Sköldenberg, O., and Gordon, M. (2017) Artificial intelligence for analyzing orthopedic trauma radiographs: deep learning algorithms—are they on par with humans for diagnosing fractures?. *Acta Orthopaedica* 88(6), 581-586.

[78]   Elkin, P. L., Schlegel, D. R., Anderson, M., Komm, J., Ficheur, G., and Bisson, L. (2018) Artificial intelligence: Bayesian versus heuristic method for diagnostic decision support. *Applied Clinical Informatics* 9(02), 432-439.

[79]   Vuong, Q. H., Ho, M. T., Vuong, T. T., La, V. P., Ho, M. T., Nghiem, K. C. P., Tran, B. X., Giang, H. H., Giang, T. V., Latkin, C., and Nguyen, H. K. T. (2019) Artificial intelligence vs. natural stupidity: Evaluating AI readiness for the Vietnamese medical information system. *Journal of Clinical Medicine* 8(2), p.168.

[80]   Emanuel, E. J. and Wachter, R. M. (2019) Artificial intelligence in health care: will the value match the hype? *JAMA* 321(23), 2281-2282.

[81]   Javaid, I. (2019) 4.0 applications in medical field: a brief review, *Curr. Med. Res. Practice* 9(3), 102.

[82]   Vavken, P., Ganal-Antonio, A. K. B., Quidde, J., Shen, F. H., Chapman, J. R., and Samartzis, D. (2015) Fundamentals of clinical outcomes assessment for spinal disorders: clinical outcome instruments and applications. *Global Spine Journal* 5(4), 329-338.

[83]   Haleem, A., Javaid, M., and Vaishya, R. (2019) Industry 4.0 and its applications in orthopaedics. *Journal of Clinical Orthopaedics & Trauma* 10(3), 615-616.

[84]   Nittari, G., Khuman, R., Baldoni, S., Pallotta, G., Battineni, G., Sirignano, A., Amenta, F., and Ricci, G. (2020) Telemedicine practice: review of the current ethical and legal challenges. *Telemedicine and e-Health* 26(12), 1427-1437.

[85]   Loomba, A., Vempati, S., Davara, N., Shravani, M., Kammari, P., Taneja, M., and Das, A.V. (2019) Use of a tablet attachment in teleophthalmology

for real-time video transmission from rural vision centers in a three-tier eye care network in India: eyeSmart cyclops. *International Journal of Telemedicine and Applications* 2019.

[86]   Molfenter, T., Brown, R., O'Neill, A., Kopetsky, E., and Toy, A. (2018) Use of telemedicine in addiction treatment: current practices and organizational implementation characteristics. *International Journal of Telemedicine and Applications* 2018.

[87]   Acharibasam, J. W. and Wynn, R. (2018) Telemental health in low-and middle-income countries: a systematic review. *International Journal of Telemedicine and Applications* 2018.

[88]   Sudas Na Ayutthaya, N., Sakunrak, I., and Dhippayom, T. (2018) Clinical outcomes of telemonitoring for patients on warfarin after discharge from hospital. *International Journal of Telemedicine and Applications* 2018.

Telehealth using Machine learning

Abstract

Clinical trial monitoring is no different from other processes in that it is vital and necessary to make sure that established protocols and procedures are rigorously followed. Given that there are human subjects involved, it is one of the most important procedures that needs to be watched. Tools from the field of ICTs, can speed up the process and get things done more quickly, increasing accuracy while trying to monitor clinical trials. In this study, a novel conceptual framework for clinical trial monitoring is developed using machine learning techniques such as deep neural networks and support vector machines to classify physiological datasets from wearable devices. The data gathering, transmission, analysis, and prediction modules make up the proposed framework prototype. The suggested framework is built with data analysis in mind, and its central component is the module for data analysis and prediction. Through various experimental analyses, these datasets are used for things like grouping together support vector machines (SVMs) and artificial neural networks (ANNs) in one bag training preprocessed and altered before being utilized for system learning and validation (ANN). The results are presented by categorizing the individuals into three groups, including fit, unfit, and unsure. To decide whether a participant should be permitted to continue the experiment or not, these several classifications are employed. This study gives a plan for monitoring clinical trials from a distance, which helps the research team make decisions.

Keywords: monitoring; machine learning; electrocardiogram; health; clinical trails

5.1 Introduction

Clinical trials test new ways to improve patient care and quality of life. Clinical trial results inspire new drugs, diagnostics, and practices. It may also spur new research that sheds light on specific disorders. Clinical trials made today's drugs and devices possible. [1]. Volunteers undertake experimental therapies under medical supervision. Clinical trials test new illness prevention, diagnosis, control, and treatment methods. Clinical trial research involves sponsors, patients, money, and researchers [1]. Clinical trials assist researchers discover if a new medical equipment, therapy, or drug is safe and effective before seeking ethical board approval. Clinical studies are open to all ages, races, sexes, and ethnicities with doctors' consent [1]. The most advanced information and communication technologies are used by several businesses, including pharmaceutical and healthcare. Technological advances can boost efficiency and productivity by improving trial administration, patient involvement, and patient stress [2]. ICT has made medical services available to people worldwide, highlighting its importance in healthcare. Patients can remotely contact doctors and caregivers using ICT. ICT lets patients get basic checkups remotely. Given the aging population, disabled people can quickly get medical care through remote monitoring. Due to the rapid development of ICT, remote patients' health can be tracked in real time, unlike the standard method of telephonically monitoring remote facilities, which could experience delays due to high call volume or technical issues, preventing patients from reaching doctors or other healthcare professionals while sick or injured [3]. Clinical trial monitoring protects participants by collecting accurate data and following research protocols and government laws [4]. Machine learning (ML) uses a training set or model to predict fresh data. Machine learning uses supervised and unsupervised methods. Unsupervised learning acquires features from unlabeled data, while supervised learning develops a model from labeled data [5]. ML methods can identify hidden patterns in complex data sets.

Machine learning can predict and judge by comparing new and old data. Machine learning algorithms automatically update with new data.

These strategies help clinicians diagnose. False alarms "Normalization" eliminates readings outside the typical range for a parameter (here, blood pressure, heart rate, body temperature, etc.). [7]. Clinical decision-making backed by mathematical models of physiology and expert opinions can improve healthcare delivery. Machine learning algorithms can reduce hospital stays and save lives [7]. SVMs and ANNs use changed data to model systems. These results are combined to build a clinical data profile for each participant's health information. Clinical profiles are utilized to determine participant cohorts and groups [6].

Wearable medical gadgets have been introduced by the development of technology. A system called the Internet of Things (IoT) links hardware, software, and physical items to allow them to communicate with one another. IoT-based technologies are required to address healthcare difficulties because of the growing population, dwindling healthcare resources, and rising medical costs. IoT has many uses in the healthcare industry, including remote monitoring, smart sensors, and medical device integration. By incorporating wearable medical gadgets that can track a patient's health in real time, the Internet of Things has made healthcare smart. These wearable devices can monitor a patient's health and provide real-time biofeedback to a hub. Mobile health uses medical and public health applications for wireless monitoring devices [8]. It remotely collects health data quickly. Doctors analyze data from these portable monitors [8].

Mobile communications and wireless technologies have made tracking trial participants and patients during interventions easier. Fitness trackers, which have sensors, may measure fundamental physiological factors like a person's pulse, temperature, arterial stress, nutritional condition, mental state, and physical activity. These wireless physical activity trackers may provide more accurate data from test subjects and volunteers, speeding patient rehabilitation [9]. Clinical trial research can use electronic technologies to recruit, administer, gather data, and retain participants [10]. Conventional approaches lengthen clinical trial research. Clinical studies should be faster with electronic technologies [10].

There are many potential participants. Clinical trial participants rarely have access to medical professionals who can explain the jargon, restrictions, and procedures. Unresolved misunderstandings may cause suspicion and hinder future medical contacts. Interventions require public access to clinical trial modalities [11]. Since clinical trial data quality monitoring is ineffective, wearables must be tracked to verify data [11].

5.2 Related works

Kartikee Uplenchwar and colleagues developed a Raspberry Pi and Arduino-based IOT health monitoring system. The system processes, transmits, and receives. The transmitting end's biosensors detect patient biopotential signals [12]. Wearable sensors track heart rate, blood pressure, body temperature, and location. Arduino and Raspberry Pi link sensors. The Raspberry Pi sends data to a predefined destination when online. Mobile phones and computers on the same network can monitor key parameters. Real-time tracking and data storage are absent.

Stephanie B. Baker et al. presented "Internet of Things for Smart Healthcare: Technologies, Challenges, and Opportunities". Disease-specific systems may use blood sugar, fall, and joint angle sensors. Network sensors feed the central node. It evaluates choices. Machine learning algorithms can find hidden patterns, recommend diagnosis and treatment, and advise clinicians on individual cases. Cloud storage should handle AI and large databases [13].

Using a Raspberry Pi 2, Kirankumar et al. created a low-cost web-based human health monitoring system. In this, the patient's health parameters like their body temperature, heart rate, blood pressure, alcohol sensor to determine whether or not they drank, ECG sensor, sound sensor, EMG sensor to determine how stressed they were, and video camera to record their live streaming video are all taken into consideration [14]. Raspberry Pi 2 microcontrollers display these settings on Putty SSL Clients. Neighborhood Wi-Fi connects Raspberry Pi to the internet. A webpage tracks the patient or newborn. The system's drawback is that the cloud cannot use the sensor data to identify a specific doctor for consultation.

Rohan Bhardwaj et al. established a data-rich healthcare system. This included electronic medical records (EMRs), which may contain structured or unstructured data. Height, weight, temperature, diagnosis, and therapy are structured health data. Machine learning can help doctors diagnose and prescribe more accurately, identify patients at risk of adverse pharmacological effects, improve patient health, and reduce costs [15].

Gurjar et al. used IoT and heart rate monitoring to detect cardiac attacks, and monitor heart rate and temperature to help doctors determine the metrics. WiFi alerts the server when these parameters exceed their maximum value. The home-monitoring gadget is portable, affordable, and prevents heart attacks. One server room worker monitors the patient and hospitals. However, real-time monitoring is impossible, data noise is a problem, and medical records cannot be saved online [16].

Raspberry Pi 3, IoT, and Pradnyavant Kalamkar created the Human Health Monitoring System. Raspberry Pi 3 links sensors and wirelessly sends sensor data to an IoT service. This data will be updated on the medical center's website. If Serene fails, the doctor receives an SMS. The system provides instantaneous information, and expedites and expands communication coverage to improve patient flexibility and quality of life. Emergency monitoring is straightforward because the hospital's symbol indicates rapid aid. This means accidental deaths decrease. However, data storage and live streaming are system drawbacks [17].

Figure 5.1: Proposed framework for monitoring clinical trials.

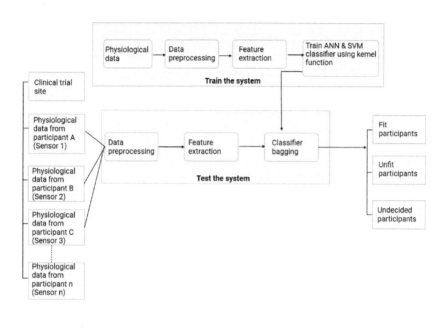

5.3 System of Watchful Waiting: A Proposed Scheme

A wearable device was used to record participants in clinical trials' measurements of their cardiovascular health, including their blood pressure, oxygen saturation, ECG, pulse rate, and respiration rate. Figure 5.1 shows one prospective treatment option for COVID-19. The suggested monitoring system receives these physiological datasets over a wireless network, where the prediction analysis is carried out [18]. The clinicians' decision-making is informed by this analysis. Data categorization and prediction analysis using techniques like support vector machines (SVMs) and artificial neural networks (ANNs), shed light on clinical trial datasets and lead to new discoveries. Clinicians have access to reviews as well as safety indications via links or graphs. Researchers and physicians can immediately access any physiological abnormalities in trial participants' data, making it simple for them to decide whether or not to keep a participant in the study based on the data analysis.

Table 5.1: The wearable device will collect the data shown in the table.

Physical indicator	Normal methods for assessing the physiological status	First-stage data collections from participants	Wearable device data from a clinical trial
Temperature (Temp)	36.5–37.5 °C (97.7–99.5 °F)	Not available (N/A)	Not available (N/A)
Heart rate (HR)	60 to 100 bpm (bits per second)	N/A	N/A
Electrocardiogram (ECG)	25 mm/s	N/A	N/A
Respiration rate (RR)	12 to 16 breaths per minute	N/A	N/A
Blood oxygen saturation (SpO2)	80 and 100 millimeters of mercury (mm Hg)	N/A	N/A
Blood pressure (BP)	120/80 millimeters of mercury ("mm or Hg")	N/A	N/A

These physiological datasets are pre-processed into a feature set and then used to train classifiers. The database is accessible via a wireless communication network and receives data from the wearable device for remote access. An emergency arises when a patient's physiological indicator reads above or below the usual in the range shown in Table 5.1.

Support vector machines and artificial neural networks can classify wearable device data by removing noise and artifacts using principal component analysis. This is SVM and ANN-based data classification [18]. People are aided in making quick decisions by outliers and likely failures. Unprocessed input is transformed into signals by these algorithms through training. Visuals and a fictitious implementation strategy are used to display signals that are easy for humans to interpret.

Data from wearable devices is pre-processed to remove missing, noisy, or inconsistent values (Biobeat). Before running the algorithm (SVM or ANN), the final training set datasets must be pre-processed. Data normalizations, data transformations, and instance selection are all involved, and irrelevant datasets should be removed. 80% of pre-processed datasets are used for training, 20% for production, and the rest for testing. Feature extraction generates a graphical representation to analyze processed datasets and creates new features from existing datasets to reduce physiological dataset characteristics.

Overfitting of SVM/ANN is avoided via feature extraction on cleaned data. This enhances machine learning and accelerates algorithm training. In noisy data, bagging lowers variance.

Classifier bagging is an ensemble strategy that uses randomization and data classification depending on trial subjects' test performance [19]. A bagging classifier ensemble meta-estimator randomly applies basic classifiers to datasets and votes or averages the results to make a single prediction. Each basic classifier uses N fewer instances than the training dataset. Each base classifier uses unique data. The training set may repeat several initial recordings.

This conceptual approach to monitoring COVID-19 patients in treatment trials is innovative and fresh due to the following:

1. The Biobeat is a unique device, and collects data for this investigation. Remote data collection from COVID-19 patients is safer [20]. The new Biobeat gadget works well.
2. The suggested wearable device can record six physiological data points, comparable to most others. Synchronization and comparison analysis issues arise whenever a researcher uses numerous devices to collect data. This study improves earlier studies by using one device to collect six body physiological data [20].
3. Most clinical studies are conducted locally; however, our study intends to improve the normal approach by remotely monitoring COVID-19 patients [21]. Biobeat remote monitoring of COVID-19 patients is not part of any scheduled clinical trials, but greatly aids COVID-19 patient monitoring.
4. Remote human health monitoring will enable rapid intervention in clinical trials with suboptimal drug outcomes for COVID-19 patients who volunteered [21].

5.4 Methodology

Arduino Uno and Raspberry Pi microcontroller boards power the system. Arduino Uno and sensors communicate serially on parameters including temperature, acceleration, ECG, pressure, and pulse sensors. Raspberry Pi Model B boards have cameras and microphones. The Raspberry Pi sends Arduino board data to DJANGO's SQL database [22]. The monitoring of websites occurs in real time. In times of crisis, SVM classifiers sort sensor data. The WAY 2 SMS service alerts doctors of data discrepancies. Emergency notifications include the ambulance service. A block diagram of a health tracking gadget is shown in Figure 5.2 [22].

Figure 5.2: Schematic representation of a possible health monitoring system.

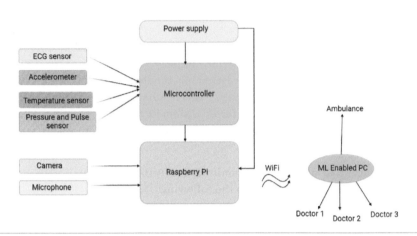

5.4.1 Support vector machine (SVM)

Support vector machines can evaluate a dataset and predict every observation [23-27]. This linear, binary classification algorithm is not probability-based. Each output type is dependent on another group. These enable automatic training algorithms to build a model that classifies new data according to predefined criteria. The hyperplane generated by support vector machines (SVMs) may not be optimal for data separation [28]. A model overfits when its results are overly particular to the training data. Errors and penalties in the model data are captured by setting a minimal gap, providing some wiggle room to correct for this kind of training error. This type of margin is called "soft margin."

Table 5.2 shows this work's SVM parameters. SVMs are a class of learning methods that predict binary qualitative variables. Later, this technique predicts

Table 5.2: SVM variable.

Parameter	Value
Kernel	Linear
Method	Sequential minimal operation (SMO)
Population number	190

Figure 5.3: Flowchart of SVM.

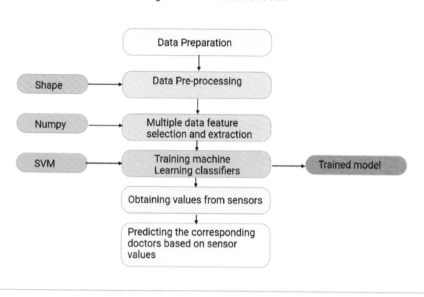

more quantitative factors. Discrimination is binary. In such cases, it seeks the hyperplane's farthest point that perfectly separates or classes the data. Thus, the goal is to find a classifier (discriminant function) with the maximum generalization capacity (prediction quality). An SVM flowchart is shown in Figure 5.3.

5.4.2 Artificial neural network (ANN)

There is a wide variety of current and potential applications for artificial neural networks [29-31] in the field of artificial intelligence; biological neurons have their mathematical counterparts in neural networks. ANNs may tackle complex issues in forecasting, optimization, pattern recognition, and more. Process objects (nodes) and their connections (functions) are the building elements of graphs in artificial neural networks (links). They are able to analyze information and find the solution to problems. Local memory is rarely implemented in models as process elements or nodes. Neural network nodes and connections are layered. Forward propagation [32] is used at the start of dataset training. The technique's main benefit is its ability to calculate and store output and intermediate variables sequentially from input to output. Thus, social classes will talk more along one axis. This approach initially sets all dataset weights

Table 5.3: ANN with front-propagation.

Parameter	Value
Variants	1000
Number of people	190
In-class instructional period	7 s

Figure 5.4: Flowchart of ANN.

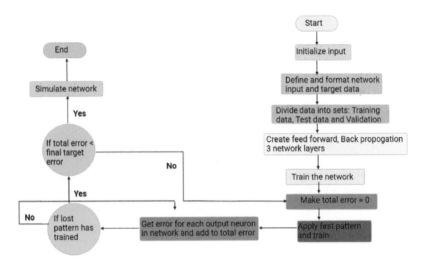

arbitrarily. An extensive analysis of all data sets produces the best projections. Table 5.3 shows the ideal parameters for a forward propagating artificial neural network (FP-ANN). The output layer organizes the hidden layer's predictions. Figure 5.4 shows ANN's flowchart.

5.5 Hardware Information

Arduino/Genuino Uno's main CPU is the ATmega328P. 14 digital I/O pins, a reset button, USB and power ports (six of which are PWM outputs). The box contains the microcontroller, ac-to-dc converter, and USB cable.

This Electrocardiogram Sensor Module Set is a complete, single-lead ECG front end that operates on 3.5 V and 170 A. The signal can be retrieved using a microcontroller or an ultra-low-noise analog-to-digital converter (ADC). Your heart's electrical activity can be tracked with a low-cost board like the SparkFun Single-Lead Heart Rate Monitor AD8232. The electrical activity of the heart can be shown graphically on an electrocardiogram (ECG). In order to quickly and easily isolate the PR and QT intervals from the noise of an ECG, the AD8232 Single Lead Heart Rate Monitor can be utilized as an operational amplifier. The serial output is used by external integrated circuits projects to display vital signs like blood pressure and pulse. It shows the user their systolic, diastolic, and pulse rates. Due to its small form factor, it can be worn on the wrist like a watch. The heart rate cannot be measured with a simple wrist watch. GY-521 MPU6050 Modules have gyroscopes and accelerometers on all three axes. The gyroscope and accelerometer on the same silicon chip can be used with Arduino. MPU-6050 chips [33]. Use the MPU-6050 chip. 3–5 V input. IIC Communication Modes are ubiquitous. It outputs 16-bit data from its onboard A/D converter. Gyroscope: 250–500–1000–2000–s. 2-16g acceleration.

The Raspberry Pi is a credit card-sized computer that works with a keyboard, mouse, and TV remote. It helps kids learn programming languages like Scratch and Python. Logitech C270 HD is easy to carry. Its 70 × 18 × 30 mm size fits comfortably in your palm. Its tilting head gives the Logitech C270 camera a wide viewing angle [33].

5.6 Description of the Program

Arduino is a free and open-source prototyping platform that uses user-friendly hardware and software to create and distribute operational prototypes. The software for all Arduino boards is free, so anyone can make their own board and alter it. It supports macOS, Windows, and Linux.

Raspberry pi uses Debian-based Raspbian, founded by Mike Thompson and Peter Green [33]. Raspbian Stretch and Jessie are available. Construction ended in June 2012, but OS development is ongoing. It's optimized for Raspberry Pi's tiny ARM CPUs.

Django, an open-source web framework, uses Python's MVC (model-view-template) architecture. This application simplifies data-driven website creation. Django emphasizes component reuse and pluggability, minimum code, low coupling, and rapid page development. Python writes all configuration files and data models [33].

5.7 Conclusion

Clinical trials are the "gold standard" for testing new medical therapies [22]. Developing wearable Internet of Things technology increases treatment efficiency and effectiveness. Wearable sensors networked with remote medical infrastructure can be widely deployed. Many people's quality of life is affected by chronic diseases including high blood pressure and diabetes. Health workers, caretakers, researchers, doctors, and others can use wearables to remotely monitor patients' and volunteers' reactions to a new medicine, improving participant quality of life [13]. In the light of the complex issues surrounding complete on-site oversight of clinical trials by government regulatory bodies and advisory ethics committees, our proposed framework provides the foundation for remote human health monitoring to mitigate any unintended consequences from COVID-19 clinical trials.

References

[1] Friedman L. M., Furberg C. D., DeMets D. L., et al. (2015) *Fundamentals of Clinical trials*. SpringerLink. https://doi. org/10.1007/978-3-319-18539-2_1

[2] Christian J., Dasgupta N., Jordan M., et al. (2018) Digital health and patient registries: today, tomorrow, and the future. In: *21st Century Patient Registries: Registries for Evaluating Patient Outcomes: A User's Guide: 3rd Edn.* Gliklich, R. E., Dreyer, N. A., Leavy, M. B., and Christian, J. B. (Eds). Rockville, MA: Agency for Healthcare Research and Quality.

[3] Su C. R., Hajiyev J., Fu C. J., et al. (2019) A novel framework for a remote patient monitoring (RPM) system with abnormality detection. *Health Policy and Technology* 8, 157–70.

[4] Hatayama T. (2020) Bayesian central statistical monitoring using finite mixture models in multicenter clinical trials. *Contemp Clin Trials Commun* 19, 1–12. https://doi.org/10.1016/j. conctc.2020.100566.

[5] Jordan M.I. and Mitchell T.M. (2015) Machine learning: Trends, perspectives, and prospects. *Science* 349, 255–60.

[6] Huang S., Cai N., Pacheco P.P., et al. (2018) Applications of support vector machine (SVM) learning in cancer genomics. *Cancer Genomics Proteomics* 15, 41–51.

[7] Rush B., Celi L. A., and Stone D. J. (2019) Applying machine learning to continuously monitored physiological data. *J Clin Monit Comput* 33, 887–93.

[8] Hassan M. K., El Desouky A. I., Elghamrawy S. M., et al. (2018) Intelligent hybrid remote patient-monitoring model with cloud-based framework for knowledge discovery. *Comput Electr Eng.* 70, 1034–48.

[9] Appelboom G., Yang A. H., Christophe B. R., et al. (2014) The promise of wearable activity sensors to define patient recovery. *J Clin Neurosci* 21, 1089–93.

[10] Rosa C., Campbell A. N., Miele G. M., et al. (2015) Using e-technologies in clinical trials. *Contemp Clin Trials* 45, 41–54.

[11] Daniel D., Kalali A., West M., et al. (2016) Data quality monitoring in clinical trials: has it been worth it? An Evaluation and Prediction of the Future by All Stakeholders. *Innovations in Clinical Neuroscience* 13, 27–33.

[12] Kartikee Uplenchwar AND Aditi Vedalankar (2017) IoT Based Health Monitoring System using Raspberry Pi and Arduino. *International Journal of Innovative Research in Computer and Communication Engineering* 5(12).

[13] Baker, S. B, Xiang, W., and Atkinson, I (2017) Internet of things for smart healthcare: technologies, challenges, and opportunities *IEEE Transactions* 5 .

[14] Kirankumar, C. K. R. and Prabhakaran M. (2017) Design and implementation of low cost web based human health monitoring system using Raspberry Pi 2. *International Conference on Electrical, Instrumentation and Communication Engineering.*

[15] Bhardwaj, R., Nambiar, A. R., and Dutta, D. (2017) A study of machine learning in healthcare. *IEEE Annual Computer Software and Applications Conference.*

[16] Gurjar, A. A. and Sarnaik, N. A. (2018) Heart attack detection by heartbeat sensing using Internet Of Things: IOT. *International Research Journal of Engineering and Technology* 5(3).

[17] Kalamkar, P., Patil, P., Bhongale, T., and Kamble, M. (2018) Human health monitoring system using IOT and Raspberry pi3. *International Research Journal of Engineering and Technology* 5(3).

[18] Enoma, D. O., Bishung, J.., Abiodun, T., Ogunlana, O., and Osamor V.C. (2022) Machine learning approaches to genome-wide association studies. *Journal of King Saud University-Science* 34(4), 1–9. https://www.sciencedirect.com/science/article/pii/S1018364722000283.

[19] Osamor, V. C. and Okezie, A. F. (2021) Enhancing the weighted voting ensemble algorithm for tuberculosis predictive diagnosis. *Sci Rep* 11(1), 1–11.

[20] Daramola, O., Nyasulu, P., Mashamba-Thompson, T., Moser, T., Broomhead, S., Hamid, A., and Osamor, V.C. (2021) Towards AI enabled multimodal diagnostics and management of COVID19 and comorbidities in resource-limited settings. *Informatics* 8(4), 1–13.

[21] Lawton, J., White, D., Rankin, D., et al. (2017) Staff experiences of closing out a clinical trial involving withdrawal of treatment: qualitative study. *Trials* 18.

[22] Sheela, K.G. and Varghese, A.R., (2020) Machine Learning based health monitoring system. *Materials Today: Proceedings* 24, 1788-1794.

[23] Alvarsson, J., Lampa, S., Schaal, W., Andersson, C., Wikberg, J. E. S., and Spjuth, O. (2016) Large-Scale ligand-based predictive modelling using support vector machines. *Journal of Cheminformatics* 8(39), 1–9. doi: 10.1186/s13321-016-0151-5.

[24] Maia, M., Imentel, P J. S., Pereira, I. S., Gondim, J., Barreto, M. E., and Ara, A. (2020) Convolutional support vector models: prediction of coronavirus disease using chest X-rays. *Information* 11, 1–19. doi: 10.3390/info11120548.

[25] Mijwil, M. M. (2021) Implementation of machine learning techniques for the classification of lung X-Ray images used to detect COVID-19 in humans. *Iraqi Journal of Science* 62(6), 2099-2109. doi: 10.24996/ijs.2021.62.6.35.

[26] Gopa, G. V. and Babu, G. R. M. (2021) An ensemble feature selection approach using hybrid kernel based SVM for network intrusion detection system. *Indonesian Journal of Electrical Engineering and Computer Science* 23(1), 558–565. doi: 10.11591/ijeecs.v23.i1.pp558-565.

[27] Said, N. S. B. M., Madzin, H., Ali, S. K., and Beng, N. S. (2021) Comparison of color-based feature extraction methods in banana leaf diseases classification using SVM and K-NN. *Indonesian Journal of Electrical Engineering and Computer Science* 24(3), 1523–1533. doi: 10.11591/ijeecs.v24.i3.pp1523-1533.

[28] Hao, P., Kung, C., Chang, C., and Ou, J. (2021) Predicting stock price trends based on financial news articles and using a novel twin support vector machine with fuzzy hyperplane. *Applied Soft Computing* 98, 106806. doi: 10.1016/j.asoc.2020.106806.

[29] Zakaria, M., Al-Shebany, M., and Sarhan, S. (2014) Artificial neural network: A brief overview. *International Journal of Engineering Research and Applications* 4(1), 7–12.

[30] Santos, D. F. and Espitia, H. E. (2020) Detection of uveal melanoma using fuzzy and neural networks classifiers. *TELKOMNIKA Telecommunication, Computing, Electronics and Control* 18(4), 2213–2223. doi: 10.12928/TELKOMNIKA.v18i4.14228.

[31] Rudra, M., Reddy, P. S., Chakraborty, R., and Sarkar, P. S. (2020) Design of frequency selective surface comprising of dipoles using artificial neural network *International Journal of Advances in Applied Sciences (IJAAS)* 9(4), 276–283. doi: 10.11591/ijaas.v9.i4.pp276-283.

[32] Narad, S. and Chavan, P. (2015) Cascade forward back-propagation neu-
      ral network based group authentication using (n, n) secret sharing
      scheme. In *Proceedings of International Conference on Information Security
      & Privacy (ICISP2015)* Elsevier, 185–191.

[33] Sheela, K. G. and Varghese, A. R. (2020) Machine learning based health
      monitoring system. *Materials Today: Proceedings* 24, 1788–1794.

Index

x